THE IMPOSTORS

THE IMPOSTORS

How Republicans Quit Governing and Seized American Politics

Steve Benen

wm
WILLIAM MORROW
An Imprint of HarperCollins*Publishers*

 For information, address HarperCollins Publishers, 195 Broadway, New York, NY 10007.

HarperCollins books may be purchased for educational, business, or sales promotional use. For information, please email the Special Markets Department at SPsales@harper collins.com.

FIRST EDITION

Designed by Nancy Singer

Library of Congress Cataloging-in-Publication Data has been applied for.

ISBN 978-0-06-302648-3

20 21 22 23 24 LSC 10 9 8 7 6 5 4 3 2 1

To my wife, Eve, and my mom, Gini

CONTENTS

THE IMPOSTORS

CHAPTER 1

"We're Not Great at the Whole Governing Thing"

Meet the Post-Policy Party

JUST TWO MONTHS INTO Donald Trump's presidency, Republican policy makers were determined to replace the Affordable Care Act with a regressive alternative. For the party, tearing down the health care reform law known as "Obamacare" was more than just a legislative priority; it was an obsession. As the Trump era began and the process moved forward, Republican leaders were practically giddy at the prospect of completing their crusade and killing their white whale.

On March 7, 2017, House Speaker Paul Ryan, a Republican from Wisconsin, showed uncharacteristic bravado about his party's prospects, telling reporters he could "guarantee" the GOP plan would pass.[1]

Seventeen days later, the effort collapsed, and there was no great mystery as to why. Republican leaders struggled mightily to craft a coherent proposal, defend its virtues, and explain exactly what problems they were trying to solve.

As his gambit unraveled, Ryan told reporters, "I think what you're seeing is, we're going through the inevitable growing pains of being an opposition party to becoming a governing party."[2]

It was an unexpected admission from one of the party's highest-profile leaders, five years removed from his stint as a vice-presidential nominee. With one unscripted comment, Congress's top GOP lawmaker had effectively conceded that the modern Republican Party, in a dominant position in the nation's capital, having controlled at least one chamber of Congress for eighteen of the previous twenty-two years, was not ready to be a "governing party."

A dejected Republican congressional staffer, reflecting on the collapse of the GOP's health care legislation, added, "I'm starting to think that while we're pretty good at winning elections, we're not great at the whole governing thing."[3]

It was more than a throwaway line from a frustrated aide—it was also a concise summary of one of the most important problems plaguing American politics in the twenty-first century.

Many voters have grown accustomed to the idea of a national competition pitting two governing parties against each other. One has a more progressive vision, the other a more conservative one, but for most Americans, Democrats and Republicans are basically mirror images of one another, each with an internal range of opinions. The electorate has long had reason to assume that both major parties were mature and responsible policy-making entities, their philosophical differences notwithstanding.

The actions of the Republican Party over the last decade have made it abundantly clear that it's time to reevaluate that assumption.

The current iteration of the GOP is indifferent to the substance of governing. It is disdainful of expertise and analysis. It is hostile toward evidence and arithmetic. It is tethered to few, if any, meaningful policy preferences. It does not know, and does not care, about how competing proposals should be crafted, scrutinized, or implemented.

The modern Republican Party has become a post-policy party.

The first GOP "repeal and replace" effort on health care was a striking example of the party failing to even pretend that substance guides its work in any significant way, but it was emblematic of a much larger truth: in recent years, Republicans have brought their post-policy posture to practically every debate and every issue.

Some of this can be explained by GOP officials being put in a position to exercise governmental power despite an ideological disposition that's reflexively antagonistic toward the nonmilitary public sector. As the *New York Times*'s Neil Irwin wrote early on in Donald Trump's presidency, most Republican officials routinely have an "aversion to doing the messy work of making policy. . . . If you make a career opposing even the basic work of making the government run, it's hard to pivot to writing major legislation."[4]

But more often than not, Republicans simply find it easier to bypass the rigors of real policy making.

Reading policy analyses, attending hearings, negotiating with rivals and stakeholders, and thinking through the consequences of policy decisions requires countless hours of tiresome and unglamorous work. Peddling poll-tested, base-motivating, half-baked, hashtag-ready talking points, on the other hand, is both painless and ideologically satisfying. For GOP officials, this doesn't seem to be an especially tough call.

If the party's candidates and officeholders were punished by voters over their disinterest toward governing, they'd have no choice but to take their official responsibilities more seriously. But Republicans' post-policy attitudes have also been embraced by many of the party's supporters—and so long as GOP candidates can win elections while being lazy about policy making, they have little incentive to change.

The implications of this approach to modern governance are all-encompassing. Voters routinely elect Republicans to the nation's most powerful offices, expecting GOP policy makers to have the technocratic wherewithal to identify problems, weigh alternative solutions, forge coalitions, accept compromises, and apply some level of governmental competence, if not expertise. The party has consistently proven those hopes misguided.

By any fair measure, the GOP excels at acquiring power and exploiting electoral structures to keep it, often in defiance of the American electorate's will. Republicans may fail in breathtaking fashion when trying to govern, but they have unrivaled expertise in gerrymandering and voter-suppression techniques.

The shortcoming quickly becomes evident after Election Day, when Republicans roll up their sleeves and clumsily try to use that power in pursuit of their ostensible priorities.

Making matters worse, while our Madisonian model of government expects the governing process to include a series of compromises—between parties, between chambers, between competing branches of government, engaged in a tug-of-war for power and influence—the GOP's shift into post-policy politics short-circuits the process itself.

One of the more common questions mainstream voters ask routinely is why the two major parties so often seem incapable of working together toward common goals. Much has been written in recent years about the asymmetric relationship between Democrats and Republicans: the degree to which the latter have become radicalized in ways the former have not, and the effects of this dynamic. As political scientists Norman Ornstein and Thomas Mann documented in their seminal work, 2012's *It's Even Worse Than It Looks: How the American Constitutional System Collided with the Politics of Extremism,* those observations are rooted in uncontested facts.

But the GOP's transition into a post-policy party represents a large piece of the same ugly mosaic. The United States' two major parties are no longer simply offering different answers to the same questions—they're now asking entirely different kinds of questions. There is no point on a common continuum at which to reach compromise.

The Democrats' approach is consistently substantive: when party officials consider a policy challenge, they tend to act deliberately, evaluating the granular details in ways Republicans rarely consider. The point isn't that Democrats are always right—they're not—but rather that they at least make a point of approaching their governing responsibilities in ways that reflect a degree of seriousness and due diligence.

Too often, when the modern Republican Party controls the levers of federal power, it decides in-depth scrutiny of major legislation is neither necessary nor desirable, since it would produce unsatisfying information the party would be inclined to ignore anyway.

The GOP's post-policy evolution—or devolution, depending on one's perspective—did not occur overnight. Early on in George W. Bush's presidency, the Republican White House launched a "faith-based initiative" intended to dramatically increase the role of religious institutions in providing social services with taxpayer funds. The president tapped John DiIulio, a University of Pennsylvania political scientist, to oversee the project.

DiIulio had high hopes for what he and his team could accomplish, though they were soon dashed. "There is no precedent in any modern White House for what is going on in this one: a complete lack of a policy apparatus," he wrote in a confidential letter in 2002.[5] "What you've got is everything, and I mean everything, being run by the political arm. It's the reign of the Mayberry Machiavellis."

The political scientist explained further that he was applying the term to White House staff, who "talked and acted as if the height of political sophistication consisted of reducing every issue to its simplest, black-and-white terms for public consumption, then steering legislative initiatives or policy proposals as far right as possible."

DiIulio's frustrations were understandable. Scrutinizing institutions from an academic perspective, he expected to see Republican officials shaping policy the way the party had traditionally operated. Throughout the modern political era, the GOP was a conservative party with core philosophical principles—free-market solutions, limited government, balanced budgets, social conservatism, and a robust national defense—which Republicans generally pursued in a rigorous and intellectually serious fashion.

But soon after arriving in the nation's capital, DiIulio grew disillusioned, dejected by a GOP administration that valued electoral goals at least as much as governing.

In the years that followed, the Republican Party's disinterest in cultivating a "policy apparatus" spread. After Barack Obama's election in 2008, the GOP's Mayberry Machiavellis took over the entirety of the party. On Capitol Hill, Republicans abandoned policy arguments altogether, rejecting the Democratic administration's ideas reflexively—and in many instances, incoherently—even when Obama agreed with his Republican rivals.

In one especially shameless instance, GOP lawmakers demanded that the Democratic White House endorse legislation to create a bipartisan commission on deficit reduction. When Obama did exactly what they requested, Republicans quickly killed the bill. In fact, six GOP senators who cosponsored the legislation ended up voting against their own proposal.

While absurd developments such as these played out in public, GOP lawmakers' offices privately stopped hiring policy staffers and started hiring media flaks, because as far as Republicans were concerned, messaging trumped governing, and selling a conservative vision to the public took priority over undergirding a conservative vision with serious legislative proposals that worked.

In May 2009, for example, GOP leaders were still licking their wounds on the heels of back-to-back election cycles in which Democrats made significant gains. Republicans were lost without a map, lacking any kind of vision or policy agenda. It was at this point that the House GOP conference chairman advised his colleagues to start getting rid of legislative staff—aides responsible for writing and scrutinizing policy proposals, giving the party its capacity to govern—and start hiring aides who would focus exclusively on media.[6]

The conference chairman at the time was a congressman from Indiana named Mike Pence. The far-right Hoosier became the nation's vice president a decade later.

Five years after Pence's staffing recommendation, in November 2013, Republican representative Pete Sessions of Texas, the chairman of the House Rules Committee, said publicly that the Republicans' House majority should give up on trying to pass meaningful legislation and instead spend the next year looking for ways to strip Democrats of their Senate majority.

"Everything we do in this body should be about messaging to win back the Senate," he said. "That's it."[7]

That statement quintessentially captured the hallmark of post-policy thinking: the belief that policy outcomes and substantive governing are largely irrelevant. It also helped encapsulate the Republican Party's posture during the Obama presidency: GOP lawmakers defined their policy preferences by reflexively opposing the White House, denying the president any legislative victories, and prioritizing the acquisition of power above all else.

For example, Kentucky's Mitch McConnell, the Republicans' Senate leader since 2007, was often candid about how he approached his responsibilities. Ahead of Barack Obama's inauguration in January 2009, the GOP senator settled on a strategy of maximalist partisanship, demanding total Republican opposition to Democratic proposals—including the ones Republicans agreed with—in the hopes of derailing the Obama presidency.[8] Whether or not this benefited the country in a time of crisis was deemed irrelevant. The United States was in the grips of the worst economic meltdown since the Great Depression, and voters had just roundly rejected his party at the ballot box, but McConnell's sole focus was on undermining a new president who'd just defeated the Republican nominee by ten million votes.

As far as McConnell was concerned, he'd cracked the code of American politics: bipartisan ideas tend to be popular because they're bipartisan. When the public sees bills pass with broad support, voters are satisfied that the parties considered a question, worked on an answer, and came to a sensible conclusion. For the Kentucky lawmaker, this meant it was necessary to say no to everything in the Obama agenda, lest anyone think the Democratic president was succeeding. "We worked very hard to keep our fingerprints off of these proposals," McConnell told *The Atlantic* magazine, referring to legislation backed by the White House.[9]

This often put McConnell and his party on the wrong side of public attitudes, but the senator believed he'd cracked that code, too. "Public opinion can change, but it is affected by what elected officials do," he told *National Journal*, a publication covering politics and policy, in March 2010.[10] "Our reaction to what [Democrats] were doing had a lot to do with how the public felt about it. Republican unity in the House and Senate has been the major contributing factor to shifting American public opinion."

In other words, the Senate GOP leader wasn't concerned with defying the will of the electorate by killing popular legislation; instead, he focused on making popular legislation less popular by trying to kill it, without regard for merit or public interests. McConnell's plan was predicated on the idea that if he could just turn every debate into a partisan food fight, voters would be repulsed; Obama's outreach to Republicans would be perceived as a failure; progressive ideas would fail; and GOP candidates would be rewarded for their obstinance.

Governing was not among McConnell's principal concerns.

The Kentucky senator added soon after, in reference to his party's approach to policy making, "The single most important thing we want to achieve is for President Obama to be a one-term president. . . . Our single biggest political goal is to give our nominee for president [in 2012] the maximum opportunity to be successful."[11]

It was an aggressively post-policy perspective, articulated in a shameless way, which helped define the era. Americans who expected powerful elected officials to be principally concerned with the public's interests were told they'd have to wait—because for McConnell, undermining the popular and democratically elected president was his party's "single most important" priority.

Had the conservative lawmaker been a campaign operative or a Republican National Committee hatchet man, the posture might have been slightly more defensible. But McConnell was in a position of public trust, elected to serve the interests of Americans and responsible for federal legislating. The Kentuckian, however, made little secret of the fact that he had other goals he considered more important.

Around the same time McConnell said his party's top priority was denying Obama a second term, Nevada's Harry Reid, the Senate Democratic leader, delivered floor remarks in which he said, "Our number one priority is still getting people back to work. And the most important change we can make is in working more productively as a unified body to help our economy regain its strength." The contrast between the party leaders helped clarify how dramatically Democrats and Republicans differed when approaching governing.

On the other side of Capitol Hill, John Boehner was similarly hostile toward Obama-era policy making during his tenure as House Speaker. In the years following the 2010 midterms, as congressional productivity dropped to levels unseen since the clerk's office started keeping track in the 1940s, the Ohio Republican, echoing his ideological predisposition, tried to characterize the data as worthy of celebration: Congress, Boehner insisted in July 2013, "ought to be judged on how many laws we repeal," rather than "how many new laws we create."[12]

When he and his Republican colleagues sought power, they told the electorate that they would work to find solutions to national problems. After having been successful at the ballot box, the House Speaker effectively tried to rebrand failure—as if GOP leaders deserved credit for their record of futility, making the clumsy transition from lawmakers to law enders.

Instead of finding solutions to ongoing challenges, Boehner was reduced to arguing that Congress should focus on undoing solutions to previous challenges. Given an opportunity to look forward and make national progress, he saw value in looking backward and rescinding what had already been done. However, Republicans weren't actually repealing laws, either. If Boehner was right, and that was the proper metric for judging his GOP conference's record, it was further evidence of the party's ineptitude.

Barack Obama's exasperation with the GOP's refusal to engage in substantive debates was hard to miss. During an October 2014 address at Northwestern University, the president marveled at Republican leaders' insistence that the nation's top economic priority was sweeping tax cuts for the wealthy, even as the gap between rich and poor grew larger.

Straying from his prepared text, Obama, visibly gobsmacked, asked his audience rhetorically, "Why? What are the facts? What is the empirical data that would justify that position? Kellogg Business School, you guys are all smart. You do all this analysis. You run the numbers. Has anybody here seen a credible argument that that is what our economy needs right now?"[13]

The questions reflected certain assumptions about how the president approached governing. In his mind, those proposing far-reaching policy changes, just as a matter of course, have a responsibility to bolster their arguments with in-depth research, not just tweets and cable-news sound bites.

Or put another way, Obama expected Republicans to approach substantive debates as if they were members of a governing party rather than a post-policy one. It's no wonder that deliberations between the Democratic president and GOP leaders invariably left participants frustrated and unsatisfied: Obama, attentive to the details of governing, encouraged Republicans to support their ideas with evidence and scholarship, while Republicans had little use for either, focused as they were solely on political and ideological goals.

It was this policy nihilism that helped open the door to a new leader who would cement the GOP's post-policy status in ways that were difficult to even imagine a few years earlier.

Despite his status as the first major-party presidential nominee in American history to have literally no experience in public service of any kind, Donald Trump was aggressively hostile to the very idea of a campaign shaped by the substance of governing and equally indifferent to learning how his own government works. His candidacy effectively served as a capstone, years in the making: a post-policy party would be led by a post-policy leader.

When Trump won anyway, his electoral incentive to be more responsible disappeared, but his governmental incentive intensified in ways he struggled to understand. Indeed, during Trump's presidential transition process—a brutal obstacle course for even the most experienced and knowledgeable of politicians—there was an expectation that the Republican would shift his attention to the arduous task of assembling an executive-branch team. The president-elect defied those expectations, instead launching a first-of-its-kind postelection tour, featuring self-indulgent, campaign-style rallies in nine red-state locales.

It was an unmistakable signal that Trump, after having ignored the importance of policy making as a candidate for a year and a half, would not change. No president-elect has time for a multistate tour during a short transition process, but the Republican made it a priority. Given a choice between grueling preparations and basking in the applause of followers, Trump had little trouble choosing the more entertaining and less substantive option.

As the tour got under way, the *New Republic*'s Alex Shephard noted, "Donald Trump, a man who has a very short attention span and requires instant gratification more or less constantly, loves campaigning because he has a very short attention span and requires instant gratification more or less constantly."[14] What the incoming president did not love was getting up to speed, preparing for the profoundly difficult job he'd sought despite knowing very little about what it entailed.

As the new administration slowly took shape, it stood to reason that Trump would surround himself with more qualified personnel whose expertise he could lean on after Inauguration Day. But between Thanksgiving and Christmas 2016, as new cabinet nominees were announced, many of the choices were as inexperienced as their new boss. Most of the incoming Trump cabinet had no governing experience, no experience in the subject area they'd oversee, no experience managing a large agency, or some combination thereof.[15]

In a dynamic that mirrored the Republican's pre-election operation, Trump had staffers whose job it had been to prepare governing plans, but as the political website Politico reported in early December 2016, Trump and his inner circle "largely ignored those plans."[16] The article added that policy experts, eager to assist the incoming White House team, found it difficult to get the attention of the president-elect and those in his immediate orbit.

The only thing that seemed to capture and keep Trump's attention during the transition process was the degree to which NBC's sketch comedy show *Saturday Night Live* made him the butt of jokes.

By Inauguration Day, Trump and his team were so unprepared for the transition of power that they were reduced to asking several dozen senior Obama administration appointees to remain in their jobs, not because Republican officials approved of their work, but because the incoming administration wasn't yet properly equipped.[17] The services of Obama's people were required for the continuity of governmental operations.

This included, among others, positions related directly to national security. On January 18, 2017, *Foreign Policy* magazine reported that Trump was poised to enter the White House "with most national security positions still vacant, after a disorganized transition that has stunned and disheartened career government officials."[18]

One career government official told the magazine, "I've never seen anything like this." That was because no one had ever seen a post-policy administration try to fill important governmental posts before.

Once Trump tried to govern from his position of policy passivity, his ignorance served as a near-constant hindrance. Not only was the president unable to persuade people to support proposals he knew nothing about—during the health care fight(s), the Republican routinely found it difficult to talk about his party's plans in complete sentences—but also Trump's vision, such as it was, became a black box that no one, including his ostensible allies, could understand.

As Trump's presidency flailed, *National Review* magazine's Rich Lowry, a conservative sympathetic to the Republican's cause, reported, "No officeholder in Washington seems to understand President Donald Trump's populism or have a cogent theory of how to effect it in practice, including the president himself. . . . Trump, for his part, has lacked the knowledge, focus, or interest to translate his populism into legislative form."[19]

History offers a long list of American presidents who failed to advance their priorities for a variety of reasons. Some couldn't persuade lawmakers to follow their lead. Others presented ideas that crumbled under scrutiny. Some saw their agendas overcome by scandal.

But Trump's post-policy posture broke new ground: he struggled because, in a rather literal sense, he had no idea what he was doing or what he was talking about. As Brian Beutler summarized in the *New Republic* around the president's hundred-day mark, "Trump has failed because he has little in the way of clear substantive objectives or strategies to meet them, and he lacks the interest in or knowledge of policy and process that he'd need to correct course. Even if he had a good ear for sound advice, and the patience required to follow it, he is surrounded by amateurs and opportunists who send him endlessly careening between contradictory goals and various tactical dead ends."[20]

In April 2017 the *New York Times* featured twenty people who served as close Trump confidants outside the West Wing. Seventeen of them had no governing experience at any level.[21] For an amateur president unfamiliar with the basics of policy making, and little interest in learning, it was hardly ideal.

The result was a president who assumed governing would be simple and who seemed baffled to discover it wasn't. "I thought it would be easier," Trump complained in 2017, referring to his months-old presidency.[22]

It was among the most believable concessions he'd ever made. Trump launched his bid for national office working from the assumption that there were straightforward solutions to every possible challenge, and every president had the capacity to accomplish great things through a combination of decisiveness and force of will. The only reason other presidents failed to follow this path to governing success was because, as far as Trump was concerned, they were all idiots. It was this foolhardy vision that led Trump to assure voters that he could "make possible every dream you've ever dreamed"[23] and "fulfill every single wish" Americans have.[24] By all appearances, the Republican's audacity was entirely sincere. He was convinced that he could take office, bring affordable and high-quality health care benefits to everyone, cut taxes, defeat our enemies, create record economic growth, fix an immigration mess, strike a peace agreement between Israelis and Palestinians, negotiate new trade deals, and eliminate the national debt—quickly and with relative ease—because the only thing standing between the United States and historic greatness was a bold leader like him who could simply bark orders and deliver results.

In reality, Trump did not know what he did not know. The nation's first amateur chief executive sought the presidency, then got elected, and then went about trying to learn what he was supposed to do with his awesome power—so long as it didn't require too much effort or interfere too much with his television-watching habits.

It was a recipe for failure for a president who found it easier to guess than to govern. Trump has replaced the traditional model of policy making with whims and impulses. After nearly two years in office, the president went so far as to boast, as if he were a real-life inspiration for Stephen Colbert's overtly buffoonish Comedy Central character, "I have a gut, and my gut tells me more sometimes than anybody else's brain can ever tell me."[25]

It was post-policy thinking at its most transparent.

When Trump tried to take data and evidence seriously, he proved himself unable to tell the difference between metrics that matter and those that don't. Well into his presidency's third year, the website Politico reported on an Oval Office meeting in which several members of Congress lobbied Trump not to withdraw U.S. troops from Syria, a shift he'd been considering, despite the advice of his administration's national security team.[26] The president ordered Dan Scavino, the White House's social-media guru, to join the conversation.

When Scavino, who had helped manage one of Trump's golf clubs before joining his boss's political operation, arrived at the meeting, the president directed his aide, "Tell them how popular my policy is." It fell to Scavino to brief the lawmakers in attendance on the online reactions to Trump's social-media content related to withdrawal from Syria.

In the process, the president made clear that when it came to deciding the smartest course in complex policy challenges, the kind of evidence that influenced him was retweet tallies—hardly the characteristic of a leader who's come to terms with the seriousness of governing a twenty-first-century global superpower.

When Obama asked Republicans five years earlier to produce "empirical data" to justify their positions, this was clearly not what the Democratic president had in mind.

At times Trump took his post-policy approach to government to otherworldly heights. In March 2018 the Republican ignored the judgments of his team and expressed public support for the creation of a militarized Space Force. Describing a conversation with White House staff, Trump explained, "You know, I was saying it the other day, because we are doing a tremendous amount of work in space. I said, 'Maybe we need a new force. We'll call it Space Force. And I was not really serious. Then I said, 'What a great idea. Maybe we'll have to do that.'"[27]

How would this differ from the existing Air Force Space Command, which was already responsible for space operations involving missile detection and satellite coordination? Why had the Trump White House balked after some in Congress proposed the creation of a U.S. Space Corps? How would a militarized Space Force overcome the restrictions created by the Outer Space Treaty, which the United States signed in 1967 and requires all signatories to use "celestial bodies" exclusively for "peaceful purposes"? The president didn't have answers to any of these questions—and by all appearances, he couldn't have cared less. The last thing Trump wanted to hear about were annoying questions on governing details that might get in the way of his fun.

By his own telling, the Space Force started as an offhand joke. Nevertheless, three months later, Trump directed the Pentagon to create a new branch of the U.S. military to turn his joke into reality—not to serve a specific policy purpose or to address a shortcoming in the existing military structure, but because the president saw totemic value in promoting something he thought sounded cool. Soon after, the White House sought $8 billion from Congress in order to pursue the dream the president stumbled onto by accident.

That money was expected to come from American taxpayers, but it wasn't the only source of revenue on the minds of the president and his team: By August 2018, Trump's reelection campaign was sending fund-raising appeals to supporters, asking them to vote on their choice for a Space Force logo. (One of the less creative alternatives was nearly identical to the official logo of the National Aeronautics and Space Administration, or NASA, but in a different color.[28] In January 2020, the agency unveiled its official choice, which looked eerily similar to the logo from the *Star Trek* franchise.[29])

It captured the president's ridiculous approach to governance: start with an unexamined joke, transition to executive directives sorely lacking in purpose or meaningful scrutiny, and end with a hashtag-ready phrase that far-right voters could chant and grab their wallets to support.

At one point, after a Thursday morning round of golf in 2018, Trump published a random tweet that read, "Space Force all the way!"[30] *The Atlantic*'s David Graham compared the president's substance-free idea to one of his most notorious and duplicitous scams: "Anyone tempted to get excited about the Space Force would be well advised to keep the Trump University precedent in mind," Graham wrote. "The seminars were a short-term success: Thousands of people signed up, creating a new revenue stream for Trump. But they eventually wised up that they weren't buying real-estate secrets so much as a bill of goods, and some of them sued him, resulting in a $25 million settlement. Eventually, misleading branding schemes have a tendency to fall back down to Earth."[31]

None of this bothered the president. In fact, he didn't even try to push back against the criticisms. Trump was too busy promoting an idea that he thought made him look forward-thinking—and using it to pry donations from his followers. He was content to leave governing details to members of governing parties.

It's easy to imagine rank-and-file GOP voters finding the post-policy critique insulting, but Republicans should be far more offended by their party than by the unflattering assessment. For those who sincerely believe in the party's stated principles, it does them no favors to elect policy makers who generally share the same vision but who ultimately don't care enough about governing to competently shape legislation or to implement conservative policies effectively once they've passed.

In other words, no one benefits from a two-party system in which one of the two no longer cares about governing responsibly. In order to recover from our stagnant status quo, Republicans will need to overcome their post-policy problem, currently affecting every area of Americans' political lives, and reclaim its status as a party capable of mature policy making.

A governing party recognizes the importance of rigorous policy making. It evaluates evidence before and after trying to implement ideas. It starts with a question and works toward an answer, rather than the other way around. It is swayed by reason, data, and a clear sense of how best to use governmental levers of power. Over the last decade, the Republican Party has failed spectacularly on each of these fronts in every major dispute of the era.

This book is not a comprehensive recitation of American politics since Barack Obama's 2008 election. Rather, it's intended as a lens through which to see political developments and an argument about the importance of their effects. After all, the art of governing and effective policy making has inherent value. Those who see serious national challenges need to be able to turn to competent public servants who are willing and able to craft and implement sensible solutions. The alternative is a system in which officials abandon their governing responsibilities, problems go unaddressed, and the public suffers from the rot.

The book is also constructed as an issue-by-issue indictment, shining a light on the defining political fights of the day, each of which featured a Republican Party that simply wasn't up to the task of governing.

It's not enough to conclude that politics in the United States is badly broken; if the system is going to recover, it is necessary to understand why. And while the problem is no doubt multifaceted, at its heart, ours is a political landscape featuring a governing party vying for power against a rival that's abandoned the pretense that public policy matters at all, creating an untenable model that's eroding the American policy-making process and failing to serve the public's interests.

The Republican Party has not always been a post-policy party, and it need not remain one indefinitely. The vital challenge facing the GOP and the civil polity at large is coming to terms with the party's collapse as a governing entity and considering what the party can do to find its policy-making footing anew.

CHAPTER 2

"Manipulate the Numbers and Game the System"

Economic Policy

THE GREAT RECESSION BEGAN in December 2007, but it wasn't until the fall of 2008 that the global crash made the economic downturn extraordinary. Americans had endured plenty of recessions; they hadn't seen anything like this since before World War II.

A national unemployment rate that had dropped to 4.4 percent in 2007 would soon reach 10 percent, fueling pain and anxiety in every part of the economy, from coast to coast. Public-opinion polls showed public fears about the economy spiking, as the major indexes on Wall Street plunged, wiping out trillions of dollars in wealth. Countless families were left wondering not only about their financial futures but also about when the United States might see economic daylight once again.

The month Barack Obama was elected—November 2008—the American economy lost 727,000 jobs. Job losses that severe over the course of a year would have reflected dreadful economic conditions, but the fact that it happened in a single month offered proof of an economy that was teetering on a cliff. A month later, during the president-elect's transition period, it lost another 704,000 jobs.

The month of the Democratic president's inauguration, while Obama was still unpacking, the economy lost an additional 783,000—a number greater than the population of the four least-populous American states. Economic growth had collapsed, global markets were in turmoil, and it fell to policy makers in the world's largest economy to address a crisis unlike anything anyone had seen since the Great Depression.

Congressional Republicans were conspicuously unprepared for the task at hand. Their confusion and policy ignorance generally fell somewhere between alarming and embarrassing.

On the sixth day of Obama's presidency, House Democrats introduced economic-stimulus legislation called the American Recovery and Reinvestment Act of 2009, or as it came to be known, the Recovery Act. The package was a three-part response to the crisis—composed of direct capital investments, tax cuts for working families, and aid to struggling states—intended to pump money into an economy that suddenly lacked demand, as consumers, investors, and businesses pulled back.

The world's largest economy found itself with a giant hole, and the Recovery Act was designed to fill it.

A week later, Republican senator John Thune of South Dakota delivered impassioned remarks on the Senate floor, summarizing the Republican opposition not just to the specific proposal but also to Democratic efforts in general to invest heavily in the American economy.[1]

To drive home his points, he focused specifically on the physical size of a $1 trillion stimulus package.

Thune was quite animated on this point, explaining in detail that if officials were to stack $100 bills on top of one another,

$1 trillion would be 689 miles high. And if they tied those bills end to end, he added, $1 trillion would wrap the circumference of the planet nearly 39 times.

To be sure, this made for striking visuals on poster board, which the senator's staff had prepared for his presentation, but Thune—in 2009, the chairman of the Senate Republican Policy Committee—wasn't making an economic argument. By all appearances, he had nothing policy focused to say about the Democratic rescue efforts, so he concentrated instead on symbolic trivia, which, in Thune's mind, was all the substance he needed.

It was emblematic of the caliber of the GOP's approach to economic analysis. On the heels of Thune's theatrics, Kentucky's Mitch McConnell, the Senate Republicans' leader, told his colleagues, "If you started the day Jesus Christ was born and spent a million every day since then, you still wouldn't have spent one trillion dollars."[2]

The Republican case against the Democratic bill had clearly taken shape: the economic stimulus intended to invest a lot of money, which meant Congress should defeat the plan, because investing in a failing economy is bad.

In the midst of a brutal crisis, with Americans counting on their elected leaders to stop the bleeding and help get the country back on track, Republicans were thrust into the most important economic debate in generations, but it quickly became obvious the GOP simply didn't know what to do, didn't care to learn, and had little interest in shaking its post-policy posture.

To a very real degree, Republicans should have looked at this as a period to enjoy a little quiet time. The nation had, after all, just concluded an era in which GOP policy makers got effectively everything they'd wanted on economic policy—from cutting taxes on the wealthy, to reducing spending on domestic programs, to reducing regulations on the nation's private sector—yet all they had to show for their efforts was a global calamity. The Republican

Party's entire approach to economic governance had been thoroughly and completely discredited.

And so, when the Great Recession began, it would have been a fine time for GOP officials to show some humility, express regret over the scope and scale of their catastrophic failures, and get out of the way. Instead, Republicans felt justified complaining bitterly—and at times, incoherently—about the ways in which the new Democratic majority was scrambling to clean up the mess George W. Bush and his GOP brethren had bequeathed.

In the early months of 2009, many pundits insisted that Obama and his party had a solemn responsibility to work with the defeated Republican minority, incorporate GOP ideas into the rescue agenda, and secure bipartisan support for any economic package. The new president made a concerted effort to do exactly that: organizing a series of policy discussions intended to secure broad backing, only to find that the Republican economic agenda amounted to little more than demanding tax breaks and opposing government investments of any kind.

The defeated GOP, stripped of power after having driven the economy into an unforgiving hole, was stuck reciting talking points that had effectively been written in crayon.

In February 2009 Indiana congressman Mike Pence, the chairman of the House Republican Conference, a top position in the chamber's GOP leadership, and the future vice president, declared at a conservative think tank, "We're advocating that Congress freeze all federal spending immediately."[3] John Boehner, two years before becoming House Speaker, made the same push, imploring Democrats to approve a bill that "freezes spending at current levels."[4]

The *New York Times*'s David Brooks, a center-right observer who was publicly skeptical of the Democrats' stimulus plan, conceded that GOP calls for a spending freeze—denying the economy capital exactly when it was needed most—were "insane."[5] It nevertheless became a staple of the Republican Party's economic vision as the Great Recession ravaged the nation.

Several other Republicans unveiled similarly absurd recovery plans of their own. Arizona congressman John Shadegg and Senator David Vitter of Louisiana unveiled what they described as a stimulus plan that cost "zero dollars" and promised to create two million new jobs "without any of your money."[6] Their big idea: opening up coastal areas for oil drilling and eliminating environmental safeguards on oil companies. How this would have created two million jobs was a detail the GOP lawmakers never got around to sharing, because when post-policy parties introduce legislative initiatives, they don't see any harm in promoting made-up numbers.

Senator Jim DeMint of South Carolina pushed an alternative to the Democratic plan that consisted entirely of trillions of dollars in tax cuts and didn't invest so much as a penny in the economy. The bill nevertheless enjoyed the support of 90 percent of the Senate Republican Conference.

"We got thirty-six out of forty-one Republican senators voting for that, which is completely crazy," the *New York Times*'s Paul Krugman, a Nobel Prize–winning economist, said soon after.[7] "How much bipartisan outreach can you have when thirty-six out of forty-one Republican senators take their marching orders from Rush Limbaugh?"

In practical terms, the columnist's question wasn't altogether rhetorical. Indeed, serious policy negotiations over economic policy quickly became pointless and, at times, exasperating. Writing in the *Washington Post* as the debate took shape, Steven Pearlstein marveled at GOP officials' lack of "economic literacy."[8] Josh Marshall of Talking Points Memo (TPM), a prominent online outlet for political news and analysis, added that "virtually everything" congressional Republicans were saying about the Recovery Act "wouldn't cut it in remedial economics."[9]

As a vote on the Democratic proposal neared, Pete Sessions, chairman of the National Republican Congressional Committee, which exists to elect GOP House candidates and protect the party's incumbents, explained the nature of his party's opposition to the rescue plan, favorably comparing the GOP's tactics to, of all things, the Taliban in Afghanistan.

"Insurgency, we understand perhaps a little bit more because of the Taliban," Sessions told *National Journal*. "And that is that they went about systematically understanding how to disrupt and change a person's entire processes."[10] The Texan added that if the Democratic majority didn't give his party more "options or opportunities," Republicans would "become an insurgency."

Many Americans wondered at the time why Democrats and Republicans couldn't put aside their partisan differences and help address a once-in-a-lifetime economic crisis. The answer was obvious: one of the two major parties was prepared to respond to the catastrophic conditions the way a mature governing party should, while the other was led in part by a member who saw value in borrowing a page from the Taliban's playbook.

Complicating matters, Democrats controlled the House and Senate, but in the upper chamber, Republicans not only opposed the economic recovery plan for reasons they strained to explain, but they also refused to allow the majority to even vote on the Recovery Act unless Democrats could overcome a legislative filibuster. Obama's party had fifty-seven votes in the hundred-member chamber at the time, but they'd need sixty in order to put the rescue plan into effect.

Left with no alternatives, Democratic leaders sat down with the only "moderate" Republican lawmakers in the chamber, who'd signaled a willingness to at least talk about a possible solution.

Eventually those efforts bore fruit: three GOP centrists ultimately agreed to support the White House's economic plan. (One of the three, Pennsylvania's Arlen Specter, became a Democrat two months later, saying he no longer recognized the Republican Party he'd joined in 1965.) But even their cooperation showed signs of post-policy tendencies: Senator Susan Collins of Maine insisted on pushing the overall price of the package below $800 billion because she considered it "a fiscally responsible number."[11]

In other words, the Maine Republican arbitrarily chose a figure she thought sounded good—based not on policy but on her personal hunch—which reduced the size of the economic boost and made the stimulus less effective than it could have been. Nevertheless, among GOP lawmakers responding to the crisis, Collins was one of the *more* responsible members of her conference.

After the bill passed, South Carolina senator Lindsey Graham, despite having railed against the plan, wasted little time arguing that his home state should benefit from federal rescue efforts. "South Carolina needs help," Graham told CNN.[12] Asked whether state officials should welcome stimulus funds, the senator added, "I think that, yes, from my point of view, you don't want to be crazy here."

In the weeks that followed, Republican "craziness" ensued: Louisiana governor Bobby Jindal signaled that his state was prepared to forgo all federal assistance from the Democratic bill. Almost immediately thereafter, South Carolina governor Mark Sanford said the same thing. Alaska's Sarah Palin, the year after seeking the vice presidency as John McCain's running mate, soon followed. It wasn't long before GOP governors Rick Perry (Texas), Haley Barbour (Mississippi), Butch Otter (Idaho), and Jim Gibbons (Nevada) all joined the same club.

Some of them were clearly looking ahead to the 2012 presidential race and the prospect of boasting to the GOP base about "rejecting Obama's stimulus." But that didn't make their posturing any less ridiculous: as their state economies drowned, an indefensible number of Republican governors were skeptical of any life preserver that came from a Democratic White House. True to form, the substance behind the posturing was nonexistent.

Each of these GOP officials ultimately took the federal funds, but they took pains to express their dissatisfaction, making sure the public realized they weren't pleased with the policy that helped get their state economies back on track.

The magnitude of the Republican Party's misjudgment was hard to overstate. GOP officials expressed great certainty that Obama's economic blueprint would fail miserably because, as far as conservative lawmakers were concerned, the combination of sweeping public investments, state aid, and middle-class tax cuts simply could not work. Their governing philosophy told them so.

"[This bill] won't work to put Americans back to work," Mike Pence argued. "It won't create jobs. The only thing it will stimulate is more government and more debt. It will probably do more harm than good."[13]

In reality, almost immediately after the Democratic president signed the Recovery Act, economic growth and hiring improved nationwide, and the Great Recession officially ended four months after Obama put the plan into action. By early 2010, the domestic economy was adding jobs again, and the longest economic recovery in U.S. history got under way. The plan proved to be one of the most important and most successful economic bills in generations: a 2011 report from the nonpartisan Congressional Budget Office, Capitol Hill's official scorekeeper, found that this one piece of legislation delivered what the nation needed when it needed it, creating up to 4.6 million jobs and adding as much as 3.1 percent to America's overall economic growth rate.[14]

Confronted with encouraging economic news, many GOP leaders decided the responsible course of action was to play make-believe. In early May 2010 the U.S. Bureau of Labor Statistics released the best monthly jobs report in four years, with roughly a quarter million jobs created the month prior. The same report pointed to the biggest jump in domestic manufacturing jobs in twelve years.

Pence responded to the data with a statement that read: "These are difficult times for America's families, and today's unemployment report delivers even more bad news. Democrats continue to advance a liberal agenda that is doing more harm than good." The real-world evidence told Republicans what they didn't want to hear, so they pretended it didn't exist.

Republicans never explained why, with the pressure on and the nation facing crisis conditions, all of their economic assessments were wrong. But just as importantly, GOP officials were similarly silent on why they never bothered to make a good-faith effort to deal with the substance of the economic debate in a serious way.

It was around this time that the party's activist base rallied under a Tea Party banner, organizing national events to protest the Democratic agenda, and demanding policymakers shift their attention to fiscal considerations. This had the effect of bringing the GOP's economic ideology of the era—focused specifically on Republican opposition to tax increases and higher deficits—into sharp relief. The Tea Party's message was, however, disjointed to the point of comedy. For example, the activists complained bitterly about taxes, despite the fact that the Democrats' economic stimulus package included billions of dollars in tax cuts, targeted specifically at the middle class.

GOP activists similarly claimed to be terrified of the effects of adding to the national debt—they would not, however, support tax increases that would have brought the budget closer to balance—though the broader fiscal circumstances quickly exposed those concerns as insincere. If Tea Partiers' motivations were pure, they would have celebrated when the deficit shrank significantly during the Obama presidency and would have howled when the deficit ballooned after he left office, watching Donald Trump ignore the fiscal priorities the Tea Party activists espoused a decade earlier.

The fact that neither of these things happened reinforced suspicions that the Republican Party and its base were motivated by bad faith. At the same time, GOP officials who may have been uncomfortable with this found it necessary to toe the line, fearing a partisan backlash in the form of far-right candidates challenging them in the primaries on the grounds that they weren't sufficiently ultraconservative.

On the heels of the Recovery Act's passage, the Obama administration also took steps toward saving the U.S. automotive industry, which was on the verge of collapse. The Democratic president wasn't prepared to let that happen, so Obama tapped funds from the Bush administration's Wall Street bailout to restructure General Motors and Chrysler—and the auto giants' balance sheets.

Republicans were apoplectic, convinced that the Democratic plan would not and could not work.[15] Representative Lamar Smith of Texas called the industry rescue "the leading edge of the Obama administration's war on capitalism," while Georgia congressman Tom Price argued similarly, "Unfortunately, this is just another sad chapter in President Obama's eager campaign to interject his administration in the private sector's business dealings."

Senator Bob Corker of Tennessee insisted that the rescue effort "should send a chill through all Americans who believe in free enterprise," and Alabama senator Richard Shelby added, in reference to GM, "It's basically going to be a government-owned, government-run company. . . . It's the road toward socialism."

The GOP's arguments were exposed as nonsensical when Obama's policy worked beautifully, salvaged the industry, and, as a bonus, turned a tidy $15 billion profit for American taxpayers, just six years after the initial investment.[16] But this was more than just an example of politicians getting a prediction wrong; it was also a political party making the mistake of evaluating a policy measure through an ideological lens rather than a substantive one.

Republicans started with an assumption—any government intervention in the private sector is a mistake—and worked backward from there. It's a staple of post-policy thinking: what works, even during a crisis, is far less important than what's ideologically satisfying. And even after the policy succeeded, GOP leaders continued to condemn it. In November 2010, long after it was obvious that the policy had proven more effective than expected, McConnell argued that U.S. automakers "should have been allowed to reorganize or fail" without the benefit of federal intervention. In June 2011, after the Democratic White House touted the fact that the Big Three automakers were turning a profit for the first time since 2004, House Speaker John Boehner's office downplayed the success of the Democratic policy as "nothing to celebrate."

By the time the 2016 presidential election cycle was under way, and GOP candidates started traveling to states where Obama's auto rescue had made a critical economic difference, Republicans were *still* complaining about the successful policy. In April 2015 Senator Marco Rubio campaigned in Michigan, where the Floridian insisted the Democratic policy wasn't the "right way" to help the industry, the results notwithstanding. Rubio added that Obama's approach was "problematic."[17]

Yes, the Democratic policy resuscitated the auto industry, the backbone of the nation's manufacturing sector. And yes, Obama's initiative saved hundreds of thousands of jobs. And yes, the entire White House enterprise demonstrated the importance of competent governmental efforts to save a struggling private-sector enterprise from collapse. But for Rubio and others in his party, the results were less important than the conservative worldview, which is reflexively hostile toward federal intervention in the private marketplace.

The debate was reminiscent of an old joke from academia: The policy works in practice, but does it work in theory?

A similar dynamic was on display in 2011, when the nation's 9 percent unemployment rate remained stubbornly high, even as month-to-month hiring had improved over its early-2009 depths. Public opinion polls showed job creation was Americans' top priority.

The newly elected House Republican majority, however, brought an unusual perspective to the economic debate. In February 2011 Speaker Boehner, relying on bogus statistics, complained about a perceived increase in public-sector hiring during the first two years of the Obama era and argued that his party's proposed budget cuts would put the country on stronger footing, even if that made the jobs crisis worse.

"If some of those jobs are lost," the Ohio congressman said, "so be it."[18]

It was bizarre to hear the nation's most powerful Republican policy maker admit that many Americans would likely lose their jobs as a result of his party's economic plans—and he didn't care. As the public hoped elected officials would make every effort to lower the unemployment rate, Boehner proceeded to push measures that would deliberately raise it higher, and he was unembarrassed about expressing his indifference to the underlying economic needs.

A few months later, the Speaker delivered a speech outlining his vision at the Economic Club of New York, a nonpartisan organization that's hosted policy discussions for over a century. Boehner marked the occasion by sounding functionally illiterate on the subject, arguing that the Recovery Act that ended the Great Recession had actually "hurt" the economy, and the country would be worse off if Congress extended the nation's borrowing limit without trillions of dollars in spending cuts.

The *Washington Post*'s Ruth Marcus scrutinized the speech and concluded that Boehner was relying on an "incoherent, impervious-to-facts economic philosophy."[19]

It was an assessment with broad applicability. How Democrats were supposed to engage in a meaningful debate with a party that had effectively declared intellectual bankruptcy was a mystery.

The Obama White House, however, didn't have the luxury of simply marveling at the GOP's ambivalence toward governing;

the economy still needed assistance. In the fall of 2011, the president delivered remarks to a joint session of Congress and unveiled a $450 billion plan to boost domestic hiring with a policy blueprint he called the American Jobs Act.

The need for the proposal was plain: the nation's unemployment rate had improved from 10 percent in October 2009 to 9 percent in September 2011, but the gradual progress was cold comfort to the millions still struggling to find work. Although the Recovery Act had turned the tide, the public was still clamoring for policy makers in the nation's capital to deliver better, faster results.

An analysis from the nonpartisan Congressional Budget Office concluded that the Democratic plan would sharply improve employment, and, thanks to a proposed modest surtax on millionaires and billionaires, the plan actually would have reduced the deficit.[20] Other independent economic analyses found that the White House's blueprint would create nearly two million jobs and shave a full percentage point off the nation's unemployment rate fairly quickly.[21]

"Any senator out there who's thinking about voting against this jobs bill needs to explain why they would oppose something that we know would improve our economic situation," President Obama said in a weekly address. "If the Republicans in Congress think they have a better plan for creating jobs right now, they should prove it."

Post-policy parties, however, neither explain nor prove much of anything.

In October 2011 Senators Lindsey Graham, John McCain of Arizona, and Rand Paul of Kentucky unveiled a Senate Republican alternative, which they called the Jobs Through Growth Act. The GOP package—a combination of unspecified spending cuts and deregulation ideas, eliminating the Affordable Care Act, and a constitutional amendment to prohibit deficits—would create five million jobs, according to its authors.

Republicans made no effort to explain how they arrived at the figure. Related questions, such as when those five million jobs would materialize, went unanswered. Asked to produce a jobs plan, the Senate GOP instead produced a substance-free wish list. McCain told reporters that he and his cohorts put together the ideas as "a response to the president saying we don't have a proposal."[22] But that wasn't much of a reason. The Arizonan all but admitted he was engaged in a political ploy, not a serious attempt at policy making, and, after seeing what the senators had come up with, it was obvious they *still* hadn't come up with a proposal—at least, not a real one. There was no depth of thought, no constructive ideas, no understanding of economic policy, no invitation for rigorous scrutiny, and, by all appearances, no access to calculators or economic textbooks.

Gus Faucher, the director of macroeconomics at Moody's Analytics, a leading economic research firm, marveled at the disconnect between what the country needed and what GOP senators proposed. "Demand is weak, businesses aren't hiring, and consumers aren't spending," he explained. "That's the cause of the current weakness—and Republican Senate proposals aren't going to address that."[23]

The GOP plan was never submitted for an independent analysis, the kind of step a governing party serious about dealing with high unemployment would have been prepared to take.

Those same Republican lawmakers rejected popular provisions of Obama's proposal, expressed no willingness to compromise, and ignored calls for up-or-down votes on the American Jobs Act's key elements. When Democrats pushed a stripped-down measure—a $35 billion proposal to save or create roughly four hundred thousand jobs for teachers, police officers, and fire fighters—the GOP killed that, too.

Asked why, Mitch McConnell said, "I certainly do approve of firefighters and police. The question is whether the federal government ought to be raising taxes on three hundred thousand small businesses in order to send money down to bail out states for whom firefighters and police work."

Either the Kentucky Republican didn't know or didn't care about the details of the plan he helped kill: the American Jobs Act called for tax cuts, not tax increases, on small businesses. The truth the Senate GOP leader was reluctant to acknowledge was that Republicans couldn't explain their opposition to Obama's plan and had no interest in presenting a credible alternative. The more the White House challenged Republicans to take the debate seriously, the more the party shrugged its collective shoulders.

McConnell never even bothered to argue that the White House's blueprint wouldn't work as intended, because as far as his party was concerned, the efficacy of the proposed legislation—in this case, the plan's ability to reduce unemployment—wasn't especially important. For a post-policy party, it never is.

In the years that followed, the Republican Party's approach to the issue did not improve. Ahead of the 2014 midterm elections, for example, House Majority Leader Eric Cantor of Virginia—shortly before he suffered a humiliating defeat in a GOP primary and left politics altogether—wrote a *National Review* article touting the Republican "policy agenda" on the economy.[24] It neglected to mention any policies.

As the *Atlanta Journal-Constitution*'s Jay Bookman explained, "Cantor attempts to disguise a mere description of the problem as a solution to the problem, and the ploy is transparent. . . . When your party's fundamental political policy is that government cannot and should not help, you can't cobble together a credible 'policy agenda' on how government can help. You just wave your hands, mumble phrases, and hope nobody notices."[25]

That assessment helped summarize a decade of GOP nihilism on economic policy making, which was guided by no discernible governing vision. Worse, GOP officials weren't just passive bystanders on economic policy. They executed a plan involving opposition to all forms of economic stimulus, fighting tooth and nail to take capital out of the economy through spending cuts; rejecting stimulative social-insurance programs such as extended unemployment benefits; undermining economic confidence through a pointless debt ceiling crisis; deliberately trying to make unemployment worse; prioritizing austerity and deficit reduction over growth; and pleading with the Federal Reserve, which is responsible for the nation's monetary policy, to raise interest rates.

In one especially memorable instance, congressional Republicans were staunchly opposed to the Democratic White House's idea of extending a payroll tax break: a policy that puts a little extra money in Americans' paychecks every month. GOP leaders insisted that Democrats finance the policy with comparable spending cuts. Paul Ryan, who'd championed a payroll tax cut in 2001,[26] when George W. Bush was chief executive, argued in 2011 that extending the tax cut for American workers "would simply exacerbate our debt problems."[27]

It was the first and only time in recent memory in which Republicans suggested tax cuts had to be paid for. It was also the first and only time in recent memory in which Republicans claimed that tax cuts add to the deficit. By all appearances, they were fighting diligently to ensure consumers had less money in their pockets, despite the fact that payroll tax breaks were traditionally a Republican idea.

Naturally, it wasn't long before a variety of observers—on Capitol Hill and off—began raising uncomfortable questions about whether Republican policy makers were deliberately taking steps to sabotage the U.S. economy, even as much of the country was still struggling with the effects of the Great Recession, for no other reason than to give themselves an electoral advantage.

In July 2011 a reporter asked Democratic senator Chuck Schumer of New York whether he believed his GOP colleagues were trying to damage the American economy on purpose. Schumer replied, "It's a thought you just don't want to believe in, because that would be [horrible]. But every day, they keep giving us more and more evidence that there's no choice but to answer the question 'yes.' They give us no choice but to come to that conclusion."[28]

In the months that followed, Schumer had plenty of company. In October Harry Reid, the Democrats' Senate leader, broached the same subject, suggesting his Republican colleagues were trying to sabotage the economy for political gain. "Republicans think that if the economy improves, it might help President Obama," the Nevadan said. "So they root for the economy to fail and oppose every effort to improve it."[29]

Republicans, at least publicly, said they found this insulting in the extreme.[30] GOP leaders insisted that their agenda was entirely sincere, offered in good faith, and reflective of the party's genuine economic analyses. The fact that its demands were at odds with common sense and practically everything we know about economic policy didn't necessarily mean Republicans were saboteurs; they were, the argument went, simply echoing conservative orthodoxy.

The defense was unbelievable. Republicans were guided by an entirely different set of economic principles during the Bush-Cheney years, 2001 to 2009, when GOP policy makers believed in increased government spending, economic stimulus, extended unemployment benefits, and larger budget deficits. It wasn't until there was a Democrat in the White House that the party decided all of a sudden that the key to prosperity was to do the opposite of everything an undergraduate would learn in Economics 101.

And, not coincidentally, once the presidency was back in Republicans' hands, as luck would have it, GOP officials switched their economic principles back, reembracing bigger deficits, increased spending, and lower interest rates. The good-faith arguments the party peddled in the Obama era evaporated the moment it became convenient to do so: January 20, 2017.

In May 2019 the *Washington Post*'s Matt O'Brien explained, "If the last few years have taught us anything, it is that it is impossible to be too cynical about what Republicans say when it comes to the economy."[31]

Noting the extraordinary timing of how GOP officials seemed to spend eight years trying to make the economy worse on purpose, only to abandon their purported conservative principles the minute one of their own became president, O'Brien added, "What a coincidence that they figured out everything they'd been saying was wrong at precisely the same moment that they took office!"

The truth was simple and unavoidable: throughout the Obama era, the Republican Party's approach to economic policy was less of an agenda and more of a political scheme to cut off Barack Obama at the knees—regardless of the consequences for the country. If that was controversial before 2017, the sabotage campaign was painfully obvious in the years afterward.

In many instances, the GOP's post-policy posturing is the result of laziness. When it came to economic policy, however, the political world witnessed something more pernicious between 2009 and 2016: an abandonment of substantive governing born of maximalist partisan tactics.

Told repeatedly and in great detail that their economic ideas were demonstrably nonsensical, GOP leaders thumbed their noses and repeated their talking points. They were unembarrassed because, when push came to shove, they simply didn't care whether their arguments had merit or not. Improving the economy was never the point of the Republicans' economic platform.

The GOP's approach went through a metamorphosis after the 2016 elections, but its disinterest in responsible governing did not. By some measures, Donald Trump managed to make a bad situation worse.

The Republican sought the presidency while assuring voters he was an economic genius who would bring his private-sector brilliance and expertise to the nation's capital. The pitch was later exposed as a sham. The *New York Times* reported in October 2018 on evidence of "dubious tax schemes" and "outright fraud" that Trump exploited to receive hundreds of millions of dollars from his father.[32]

The findings painted a picture in which the president, far from the self-made man he pretended to be, relied heavily on legally dubious handouts. It was brutal, not just because of the evidence of financial wrongdoing, but also because it exposed Trump, with an embarrassing record of bankruptcies and failures to repay bank loans, as a fraud. The self-aggrandizing myth of the New Yorker as a financial virtuoso was shattered.

The *Times* advanced the story in May 2019 in ways that made matters even worse for Trump.[33] As his public profile was growing in the mid-1980s, and he paid Tony Schwartz to ghostwrite his best-selling book *The Art of the Deal*, Trump was losing money at a breathtaking pace. The newspaper found that during this period, Trump appeared to have "lost more money than nearly any other individual American taxpayer."

Or put another way, Trump wasn't just a loser in business, he was among the biggest losers in the United States.

It therefore shouldn't have been too big of a surprise when he became president and struggled to understand economic policy at an even rudimentary level. Just a few months into his presidency, Trump sat down with the weekly magazine *The Economist* to discuss his vision on all things economic and proceeded to make a series of claims that showed him to be utterly clueless.[34]

During a relatively brief interview, Trump complained about a trade deficit with Canada that did not exist; whined that the United States was "the highest taxed nation in the world," when we were actually among the lowest; and insisted that Ronald Reagan's 1986 tax plan increased the deficit, when it had been specifically designed to be revenue neutral.

At that point in the conversation, *The Economist* asked whether his party's tax plan might increase the deficit, and the president responded by suggesting he invented a familiar fiscal-stimulus metaphor.

"You understand the expression 'Prime the pump'?" Trump asked. When the journalists in the room said they were, of course, familiar with the phrase, he added, "Have you heard that expression used before? Because I haven't heard it. I mean, I just . . . I came up with it a couple of days ago, and I thought it was good."

It fell to dictionary publisher Merriam-Webster to note soon after: "The phrase 'priming the pump' dates to the early nineteenth century."[35] But putting that aside, it didn't even apply to the Republican's approach to economic policy: "priming the pump" refers to temporary stimulus intended to stir activity in a stagnant economy; Trump was pushing for permanent tax cuts for an economy that was already healthy. The president had tried to sell the *The Economist* on the idea that he'd invented an old metaphor he didn't actually understand.

In the months that followed, the Trump-era debate over economic policy deteriorated further, fueled by a president who couldn't be bothered to get up to speed on the basics. Indeed, as the Republican's presidency entered its third year, Gary Cohn, the former president of Goldman Sachs and the former top economic voice in Trump's White House, was asked by MSNBC anchor and correspondent Stephanie Ruhle whether the president understood "basic economics." Cohn, who seemed to find it extremely frustrating trying to advise Trump on economic issues during his fourteen-month tenure, refused to answer—which was itself telling.[36]

Soon after, former Federal Reserve chair Janet Yellen participated in a radio interview and was less diplomatic. Asked if she thought the president had "a grasp of economic policy," Yellen said plainly, "No, I do not."[37]

Trump spent much of his tenure proving Yellen correct. For example, the forty-fifth president wrestled with how to read reports on the nation's gross domestic product, a metric that points to the rate at which the overall economy grows (or in some cases, shrinks). Six months into his presidency, the U.S. Commerce Department reported growth of 2.6 percent between April 2017 and June 2017. Trump was eager to celebrate. He declared at a cabinet meeting, in reference to the GDP figures, "2.6 is a number that nobody thought they'd see for a long period of time. . . . And 2.6 is an unbelievable number."

That didn't make any sense. The 2.6 percent quarterly growth rate was very much in line with expectations, and it offered evidence of a healthy economy growing steadily, but it was also quite believable. Two years earlier, when Trump was in the process of kicking off his presidential candidacy and telling Americans how dreadful the economy was, there was quarterly GDP growth of 3 percent. In fact, throughout the Obama era, once the Great Recession had ended, there were plenty of individual quarters in which growth topped 4 percent.

But Trump's economic literacy was so poor, he struggled to understand the data. At a rally in Phoenix in August 2017, he boasted, "Economic growth has surged to 2.6 percent. Remember, everybody said, 'You won't bring it up to 1 percent. You won't bring it up to 1.2 percent.'"

No one had made any such claims. The president was under the impression that he'd somehow defied insurmountable odds, but economic growth was actually quite tepid in his first year in office, and while the GDP improved in Trump's second year, the numbers in 2018 fell short of the growth Americans had seen in 2015—and they never reached the levels Trump promised to deliver during his candidacy.

By October 2017, when the data pointed to GDP growth of 3.1 percent, he told *Forbes* magazine, "Most of the folks that are in your business, and elsewhere, were saying that would not be hit for a long time. You know, Obama never hit the number."[38]

In reality, quarterly economic growth topped 3 percent in 2009, 2010, 2011, 2013, 2014, and 2015. Trump evidently didn't know what he was talking about, and, as is seemingly true of all post-policy politicians, he wasn't willing to make any effort to learn.

In 2018 Trump went so far as to claim, "The GDP since I've taken over has doubled and tripled."[39] It wasn't clear if the president thought he'd actually doubled and tripled the size of the American economy, or whether he thought he'd doubled and tripled growth rates, but either way, the rhetoric was demonstrably ridiculous. It was the kind of statement one might expect from someone confused about the meaning of "GDP," "doubled," and "tripled."

As a presidential candidate, Trump vowed to deliver annual economic growth more robust than anything Americans had seen since the end of World War II. A vote for Trump, he said in 2016, would be a vote for 4 percent GDP growth.[40] At one point, the Republican added that he saw "no reason" why 6 percent growth was out of reach.[41]

In true post-policy fashion, Trump pulled those numbers out of thin air because he thought they sounded impressive. There was no policy analysis, no number crunching, and no accounting. The result was a substantive mess as well as a political one: not only did Trump peddle economic claims with little basis in reality, driven largely by his indifference toward details, but also he ended up setting unrealistic benchmarks he couldn't reach while in office.

It was throughout this period that Trump told the public, ad nauseum, that he was single-handedly responsible for creating the most robust economic miracle the world had ever seen. Reality betrayed him.

While economic growth rates underwhelmed, the president's jobs record told an even less impressive story: U.S. job growth in Trump's first year fell to a six-year low, and across his first three years in the White House, fewer jobs were created than in the comparable period preceding his tenure.[42]

The week before his inauguration, Trump promised, "I will

be the greatest jobs producer that God ever created." He failed to deliver, at least in part because he didn't know how.

At the heart of the president's economic agenda were two core elements: tax cuts and trade. Both proved to be among Trump's more embarrassing failures, driven by his inability to govern effectively on either issue.

There was a brief GOP flirtation with tackling tax reform in 2013. At the start of every Congress, the majority party's leadership saves special bill numbers as a way of designating its top priorities. H.R. 1 through H.R. 10 are set aside for the ten bills the majority party is most eager to pass, and in the 113th Congress, Republicans gave the special H.R. 1 designation to a dramatic overhaul of the nation's tax code. (*H.R.* denotes a bill that originated in the House of Representatives; *S.*, a bill introduced in the Senate.)

The party lost interest soon after. In November 2013 Politico reported that GOP members of the House Ways and Means Committee, which would have been principally responsible for crafting the legislation, wanted to devote all of their time to attacking the Affordable Care Act, not reforming tax law.[43] Congressional Republicans saw their ostensible governing goals as a distraction from their political agenda.

Given a choice between complaining about Obamacare and pursuing their top legislative priority, GOP lawmakers had few qualms selecting the former. After all, the latter would have involved an enormous amount of hard work and difficult choices, both of which the party was eager to avoid.

At a Capitol Hill press conference in February 2014, Speaker John Boehner was peppered with questions about the details of tax reform. Largely uninterested in the specifics of the prospective legislation, the Ohio Republican told reporters, "Blah-blah-blah-blah." If a party's post-policy attitudes can be summarized in one dismissive four-word phrase, Boehner did the job nicely.

The Trump-era tax reform push was marginally more constructive, at least insofar as Republicans passed a bill, but the seriousness with which GOP policy makers approached the debate was every bit as galling.

As the debate got under way in earnest, there was a fairly broad consensus on what Americans hoped to see in the Republicans' tax plan. In April 2017 the Pew Research Center, a prominent nonpartisan think tank, published a detailed report on public attitudes related to tax policy. It found that 60 percent of Americans were bothered that the wealthy weren't paying their fair share.[44] A slightly larger percentage of the public was concerned that big corporations were not paying enough either.

The public's positions didn't change as the process plodded forward. According to a September 2017 NBC News–*Wall Street Journal* poll, most Americans backed tax increases on both the wealthy and large corporations[45]—nearly identical to a *Washington Post*–ABC News poll from the week before.[46] GOP officials saw this data, rejected the public's concerns, and pursued a tax plan predicated on the idea that the richest Americans and big corporations were in dire need of a giant tax cut.

In the spring of 2017, the White House unveiled a document purporting to show Team Trump's tax plan. It was roughly five hundred words—about the length of a small brochure—featuring vague text filling just a single sheet of paper. To describe it as a policy blueprint would have stripped the word of any meaning.

After Treasury Secretary Steven Mnuchin and National Economic Council Director Gary Cohn unveiled the sheet of paper, many of the tax experts hoping to scrutinize relevant angles of the plan—how much it would cost, who would benefit, and so forth—realized quickly there wasn't enough meat on the bones to do any real analysis.

We soon learned why: the president had seen a *New York Times* op-ed written by four conservative voices who'd advised Trump on economic matters during the 2016 campaign. Steve Forbes,

Larry Kudlow, Arthur Laffer, and Stephen Moore presented a vague far-right vision on tax reform in their opinion piece.[47] Only one of the four, Laffer, could be fairly described as an economist, and his claim to fame was the discredited "Laffer curve," which falsely asserts that tax cuts can pay for themselves through increased economic growth. (Kudlow, then a TV commentator on CNBC, replaced Cohn as NEC director in 2018, while Moore was later chosen for a seat on the Federal Reserve before having to withdraw when his record of previous comments about monetary policy created an untenable controversy.)

Two days after the op-ed was published, the president bragged to the Associated Press, "I shouldn't tell you this, but we're going to be announcing, probably on Wednesday, tax reform." No one at the White House had any idea what Trump was talking about, but Politico reported that the president summoned his staff and told aides to make the op-ed his new tax plan.[48]

The Forbes-Kudlow-Laffer-Moore opinion piece urged policy makers to follow a predictable course: slash rates for the wealthy and disregard the impact on the nation's finances. Congressional GOP leaders had a different course in mind, but the president neither knew nor cared about their preferences. Trump was equally inclined, on a whim, to disregard the work his own team had previously conducted on formulating a blueprint for the administration.

The result was an official White House tax plan that read like it had been thrown together by a distracted intern the morning of its unveiling.

In the months that followed, Republicans sketched out a vision of why the issue was a top priority for the party: GOP leaders described a tax system that was overly complex, littered with loopholes, and burdened by too many brackets. The party believed it could simplify the system, while delivering new benefits to millions of middle-class families.

Speaker Paul Ryan placed a special emphasis on reforming the tax code so that Americans, frustrated by the annual burden of filing out byzantine tax materials, could file their returns on a postcard.

In the abstract, there was nothing especially offensive about the pitch. But in practice, Republicans proceeded to do largely the opposite of what they claimed they were determined to do.

The extent to which the GOP tax plan was a betrayal has been well documented. Donald Trump asserted in his inaugural address, "The forgotten men and women of our country will be forgotten no longer," and less than a year later, he made a giant corporate tax break the centerpiece of a tax package that largely ignored the needs of working families. The entire initiative represented a radical wealth-redistribution experiment, directing funds to those who least needed a federal rescue during healthy economic times.

But viewing the tax plan fight through a post-policy lens exposed a parallel problem: the Republican plan wasn't just a betrayal but also a dreadful piece of legislation considered and approved in such a way that proved just how little GOP policy makers cared about the substance of governing.

In 1986, when the Reagan White House and a divided Congress took up the issue of tax reform, policy makers held more than a dozen legislative hearings with experts and invested six months into detailed negotiations.[49] The result was a bipartisan bill that succeeded in what it set out to do.

In 2017, GOP officials wrote a bill behind closed doors, deliberately excluded Democrats, and ignored every analysis submitted by independent scorekeepers and subject-matter experts. House Republicans unveiled a bill in early November 2017 and moved it through committee in a matter of days without so much as a single hearing.

New York Times columnist Gail Collins, mocking GOP lawmakers as part of an imagined Q&A, wrote, "That's really sweet of you, imagining there'd be hearings. Congress doesn't have public deliberation while it's preparing big bills anymore. It's totally not Age of Trump."[50]

Republican Tom Cole of Oklahoma, a prominent member of the House Appropriations Committee, was asked why he supported a bill he knew little about, rejecting the advice of economists and budget analysts who urged Congress not to advance a plan that would blow a hole in the federal budget. "In the end, I'm going to trust the people who are philosophically aligned with me," the Oklahoman said, rationalizing his lack of concern about the lawmaking process.[51]

In a comment that seemed to capture the attitudes of too many of his GOP colleagues, Cole added, "I can't tell you I'm a deep economic thinker."

On the heels of the bill passing the lower chamber, Mick Mulvaney, the White House budget director, made multiple appearances on the Sunday news shows, where he all but admitted that his party wasn't even trying to take the governing process seriously. "A lot of this is a gimmick," Mulvaney said on NBC, in reference to the plan's financing.[52] On CNN, he added, "It's simply trying to essentially manipulate the numbers and game the system."[53]

The *Washington Post*'s Ruth Marcus wrote in response, "In other words, we're lying to you to ram this through, and we're not even going to bother to hide it." Recalling English satirist Jonathan Swift's observation—"I never wonder to see men wicked, but I often wonder to see them not ashamed"—Marcus added, "If hypocrisy is the tribute that vice pays to virtue, what does it say, exactly, when our most senior public officials feel no such compunction? What does it mean if we lose Swift's capacity to wonder at the absence of shame?"[54]

It was against this backdrop that the process in the Senate managed to be even more cringeworthy. In late November 2017, Senate Republicans voted on a procedural measure to bring their tax plan to the floor, skipping both legislative hearings and the part of the process in which they were supposed to understand their bill before voting on it. Lawmakers conceded that they couldn't read the chamber's version of the plan because they hadn't yet written it—even as they prepared to pass it.

Senator Chris Murphy of Connecticut, a Democrat, marveled a day later, "Uhhh, we're hours away from a multitrillion-dollar rewrite of the tax code, and the people voting for it don't know what's in it."[55] GOP senators were effectively legislating in the dark, wearing a self-imposed blindfold.

It would have been easy for responsible lawmakers to tap the breaks, take a breath, and recommend a detailed hearing or two about exactly what was on the table. But Republicans weren't willing to consider even such a small step.

On December 1, 2017, draft legislation was circulated featuring hard-to-decipher text written by hand in the margins. Democratic senator Jon Tester of Montana, visibly amazed, complained, "I was just handed a four-hundred-seventy-nine-page tax bill a few hours before the vote. One page literally has hand-scribbled policy changes on it that can't be read. This is Washington, DC, at its worst."[56] Illinois senator Dick Durbin, a fellow Democrat, described the GOP scribbles in the margins of the bill as "artwork," a word rarely associated with tax legislation.

Business Insider published the memorable headline "Something Very Stupid Is Happening in the Senate Right Now."[57]

The plan passed anyway, with the unanimous support of every Republican in the Senate, including the alleged moderates. Tennessee's Bob Corker, who'd vowed publicly to reject the legislation if it added "one penny to the deficit," reversed course at the last minute and voted for the bill, disregarding the overwhelming, independent evidence that the tax breaks would add $1 trillion—or a hundred trillion pennies—to the national credit card.

The GOP's entire endeavor on tax policy, from start to finish, wrapped up in six and a half weeks. At the White House bill-signing celebration, the president hailed the lawmakers who "worked so long" and "so hard" on the legislation.[58] It was the only unintentionally amusing moment of the process.

Within a year of the law's passage, practically every relevant metric pointed in the same direction: the Republicans' tax plan was a failure. The breaks didn't fuel private-sector hiring,[59] didn't boost business investment,[60] and failed spectacularly to pay for themselves.[61] What's more, many corporations used their windfall for stock buybacks, boosting their prices on Wall Street indexes, rather than improving workers' salaries or expanding their workforce.

It was a political flop, too. In late 2017, as the regressive tax plan was poised to clear Congress, Mitch McConnell boasted, "If we can't sell this to the American people, we ought to go into another line of work."[62] The policy's popularity never grew, and as the 2018 midterm elections approached, Republicans largely omitted their only major legislative accomplishment of the Trump administration's first two years from their campaign messaging.

In April 2018 Donald Trump was supposed to deliver remarks on his party's policy at a tax roundtable in West Virginia, as part of a larger election-year public relations campaign in support of the unpopular law, but he couldn't be bothered. "You know, this was going to be my remarks, it would've have taken about two minutes, but to hell with it," the president said, literally throwing papers in the air. "That would have been a little boring, a little boring."[63] He used the tax roundtable to attack immigrants instead.

But the larger embarrassment was the degree to which the plan also failed by the GOP's own terms: it hadn't simplified the tax code, the number of tax brackets remained unchanged, and middle-class families were not among the plan's principal beneficiaries. The chasm between what Republicans said they considered important and what they produced was so significant, there was a hearty debate about what could explain such rampant incompetence.

As for Paul Ryan's postcard test, the Trump administration unveiled a postcard-sized form, but as Politico reported in July 2018, "The move was nothing more than a publicity stunt—as a number of commentators noted, the administration achieved its postcard-sized ambitions only by requiring millions of Americans to submit supplementary worksheets that actually complicate the task of tax preparation."[64]

In case that wasn't quite enough, the Republican plan was littered with a series of technical errors and glitches—including, in one instance, a $250 billion mistake with the corporate alternative minimum tax, which senators scrambled to address after they'd begun voting on it.[65] Once the tax breaks were implemented, a series of related problems emerged, adversely affecting houses of worship,[66] Gold Star families,[67] and low-income college students[68] who rely on financial aid, among others.

While some slipups are unavoidable in major pieces of legislation, Marty Sullivan, chief economist at Tax Analysts, a nonprofit firm focusing on tax-policy research, said in February 2018, "This is not normal. There's always this kind of stuff, but the order of magnitude is entirely different."[69] Most of the errors likely would have come to light far sooner if Republicans in Congress had taken the time to scrutinize their own plan through legislative hearings. They didn't bother, uninterested in the real-world implications of their handiwork.

It didn't have to be this way. Subject-matter experts spent months trying to warn GOP lawmakers and present them with evidence that their plan was, among other things, wildly unnecessary. Corporate profits were already at record highs; the wealthy already had ample disposal income; and adding $1 trillion to the deficit when the economy was healthy defied Economics 101.

Republicans even had a case study to examine. In Kansas, Republican governor Sam Brownback and a Republican-dominated legislature had already approved a similar tax package years earlier. Soon after, the state's job growth and economic growth rates lagged behind neighboring states, its budget was in shambles, and Kansas's debt rating was downgraded multiple times.

And yet, in the nation's capital, the post-policy party was unmoved.

To govern is to choose, and crafting a credible tax reform package is supposed to force policy makers to make all kinds of tough choices. The trade-offs, such as identifying which loopholes to close and which tax breaks to eliminate, are always the hardest part. Republican policy makers, however, embraced a strategy in which tough choices were treated like policy analyses: annoying distractions that needlessly interfered with the task GOP officials had already decided to complete. The more that economists, tax-law experts, and scorekeepers at the Joint Committee on Taxation, which includes experts who focus exclusively on tax proposals, and the Congressional Budget Office said the party's numbers didn't add up, the more GOP leaders urged everyone involved in the process to ignore the eggheads with calculators.

At one point, the University of Chicago Booth School of Business surveyed a group of leading economists, asking whether they agreed with the Republican contention that their tax breaks would pay for themselves. Only two of the thirty-seven respondents endorsed the idea that the policy would create enough growth to cover its costs.[70]

The two Chicago economists sympathetic to the GOP's pitch clarified soon after that they'd misunderstood the questionnaire and they actually disagreed with the Republicans' argument, bringing unanimity to the scholars.

Republicans had an opportunity to pursue their own goals in a more responsible way, but, in short order, they presented unsettling proof that they didn't know how. GOP officials managed to pass a bill but they'd nevertheless failed to govern like adults.

For his part, Donald Trump was eager to take credit for his party's unpopular tax plan, though there was never any evidence of him doing substantive work on the legislation. The president seemed far more interested in branding considerations, urging GOP lawmakers to call the bill the "Cut Cut Cut Act."[71]

After the tax plan became law, the *Washington Post* highlighted some behind-the-scenes details of Trump's role in the process, including the fact that the president "always wanted the individual rates to be multiples of 5"—not for any substantive reason, but because the president simply thought those numbers sounded nicer.[72] The report added that Trump had opinions on which numbers he considered "pretty."

After signing the package into law, the president climbed aboard Air Force One and headed to Mar-a-Lago, his private resort in Palm Beach, Florida. That evening, Trump spent some time with his exceedingly wealthy patrons, telling them, "You all just got a lot richer." The president, his White House team, and his congressional allies had spent the better part of a year telling the public that the plan wouldn't benefit the richest Americans.

"The people I care most about are the middle-income people in this country, who have gotten screwed," Trump told the *Wall Street Journal* in July 2017.[73] "And if there's upward revision, it's going to be on high-income people." In the same interview, the president said he'd spoken to billionaire Robert Kraft, owner of the New England Patriots football team, who told him, "Donald, don't worry about the rich people. Tax the rich people. You got to take care of the people in the country."

Trump added, "It was a very interesting statement. I feel the same way." If he intended to act on those feelings, he failed in dramatic fashion.

The president's approach to trade policy has hardly been better. Few areas animate Trump as much as trade deals, and for many years, he has described this as his signature issue—which was unfortunate, given his disinterest in learning the basics.

After leaving his role as the White House's chief strategist just seven months into the Trump presidency, Steve Bannon spoke with PBS's *Frontline* about his first meeting with Trump and their joint interest in international trade. "He didn't know a lot of details," the always loquacious Bannon revealed. "He knew almost no policy." Bannon added that "a lot" of what Trump said was little more than recycled material "he'd heard from Lou Dobbs," a conservative television personality on Fox Business Network.[74]

Three days after his inauguration, Trump nevertheless got to work on his trade agenda, announcing the formal demise of the Trans-Pacific Partnership (TPP), a 2016 agreement endorsed by the United States and eleven other countries, which the Republican president claimed to hate for reasons he struggled to explain. By all appearances, Trump had no idea what the TPP was or what it entailed, but he was nevertheless certain he didn't like it.

About a year earlier, during a Republican presidential primary debate, a moderator asked Trump to explain his opposition to the proposed agreement. The candidate rambled on about China and currency manipulation. It fell to another White House aspirant, Senator Rand Paul, whom no one would describe as a policy wonk, to interject, "You know, we might want to point out China is not part of this deal."

It was a detail Trump seemed completely unaware of. It was also the first indication that he preferred to reflexively denounce any and all trade agreements that he wasn't involved in negotiating. If that meant rejecting sound policies sight unseen, so be it.

After the TPP's official demise, the president assured Americans that he'd replace the international compact with a "beautiful" alternative. In the years that followed, Trump never bothered to try. In fact, after the White House's decision in early 2017, the eleven other partnering countries negotiated a trade pact of their own. It closely resembled the original TPP, except that going forward, the United States wouldn't be a party to the agreement, and the provisions the Obama administration had negotiated to benefit American businesses were gone.

Two years later, at the urging of two dozen GOP senators, Trump briefly considered reversing course on the Trans-Pacific Partnership—before abruptly slamming the door shut days later. Larry Kudlow, in his capacity as director of the White House National Economic Council, whose members are responsible for helping advise the president on economic policy, conceded that the idea of the United States rejoining the TPP was more of a "thought than a policy."[75]

Unfortunately, with a post-policy president, that same phrase could be applied to practically every aspect of the White House agenda: Trump has always found it easy to come up with thoughts, but actual policies require a skill set the president consistently seems to lack.

Trump complained later that the United States' former TPP partners moved on without us—which sounded a bit like a guy complaining about a former romantic partner dating other people after a breakup he'd initiated.

The president's interest in negotiating a series of his own trade deals was intense, but often derailed by his confusion. A few months after taking office, Trump welcomed German chancellor Angela Merkel to the White House and proceeded to ask her over and over again about the prospect of a bilateral agreement between their two countries. In each instance, she had to remind her host that Germany belonged to the European Union; therefore, any trade deal would have to be made with the EU, not between the United States and individual European countries.

"Ten times Trump asked [Merkel] if he could negotiate a trade deal with Germany," a senior German official told the *Times of London*.[76] "Every time, she replied, 'You can't do a trade deal with Germany, only the EU.' On the eleventh refusal, Trump finally got the message. 'Oh, we'll do a deal with Europe, then.'" The same reporting added that Merkel told leading German officials that Trump had "very basic misunderstandings" on the "fundamentals" of the EU and trade.

As the Republican's presidency unfolded, Trump's "misunderstandings" on trade were not limited to Europe. In March 2018 the president spoke at a fund-raiser in Missouri, where he insisted, among other things, that Japan utilizes underhanded gimmicks to deny its consumers access to American cars. "It's called the bowling-ball test," he claimed. "Do you know what that is? That's where they take a bowling ball from twenty feet up in the air, and they drop it on the hood of the car," the president said of Japan. "And if the hood dents, then the car doesn't qualify. Well, guess what, the roof dented a little bit, and they said, 'Nope, this car doesn't qualify.' It's horrible, the way we're treated. It's horrible."[77]

No one had any idea what in the world Trump was talking about, and the bowling-ball test was a figment of his imagination. That said, Japan had imposed some barriers on American auto imports, which, ironically, it had agreed to lower as part of the TPP negotiations with the Obama administration.

At the same fund-raiser, Trump described a conversation he'd had with Canadian prime minister Justin Trudeau about trade deficits. According to the president's version of events, the Canadian leader was adamant that the United States does not have a trade deficit with Canada, while Trump was equally adamant of the opposite.

"I said, 'Wrong, Justin, you do.' I didn't even know. . . . I had no idea," he told donors. "I just said, 'You're wrong.' You know why? Because we're so stupid. . . . I sent one of our guys out, his guy, my guy, they went out, I said, 'Check, because I can't believe it.'"

Trump's bafflement notwithstanding, according to his administration's own data, the United States enjoys a trade *surplus* with Canada. But what made the anecdote so striking was the president's willingness to describe his own country as "stupid," while bragging about confronting an allied leader over trade details that, by his own admission, he had no working understanding of and which ultimately proved incorrect.

For most presidents, this would have been a deeply embarrassing moment that he hoped the public would never find out about. For a post-policy president, unconcerned about substantive details, it's an anecdote to be repeated with pride.

Trump's conversation with Trudeau had a specific purpose: the Republican had made renegotiating the existing North American Free Trade Agreement (NAFTA) a high priority for his administration. In a dynamic reminiscent of the TPP, the president's frequent condemnations of the trilateral deal were so vague, it was never clear if Trump knew anything about NAFTA beyond its name, its participants, and the fact that he hadn't negotiated its terms.

In the fall of 2018, negotiators from the three NAFTA nations—the United States, Canada, and Mexico—completed their work on a revamped NAFTA, though the overhaul was far from dramatic. Trump bragged, "It's not NAFTA redone; it's a brand-new deal," but whether he understood this or not, the basic framework of the new deal seemed familiar for a reason. Writing in the *New Republic,* Canadian journalist Jeet Heer explained that the new agreement was "only a minor shift from the status quo," adding that the changes were "mostly cosmetic."[78]

The *New York Times*'s Paul Krugman added, "My original prediction on Trump/NAFTA was that we would end up making some minor changes to the agreement, Trump would declare victory, and we'd move on. That's what seems to have happened."[79]

This model became a staple of Trump's presidency: the Republican's approach to practically every trade deal was to tweak what was already in place and then crow that he'd defied the odds and completed a task of extraordinary historic significance. Indeed, around the same time as the new NAFTA, the Trump administration agreed to some minor adjustments to a trade agreement with South Korea. The president boasted, "I'm very excited about our new trade agreement. And this is a brand-new agreement. This is not an old one, rewritten. This is a brand-new agreement." It was not a brand-new agreement; it was the old one, with minor modifications.

When it came to the talks among the North American neighbors, Trump's area of interest was the trade deal's name. The *Wall Street Journal* reported[80] in September 2018 that the U.S. president was principally concerned not with the agreement's core provisions but, rather, with "rebranding" NAFTA, and the administration's delegation stipulated that a name change for the pact was "nonnegotiable."[81]

Negotiators ultimately agreed—NAFTA 2.0 was labeled the U.S.-Mexico-Canada Agreement, or USMCA—but the American president's skewed priorities were emblematic of his post-policy posture. Soon after, it became clear that many of the foundational elements of the new NAFTA were recycled from Obama-era negotiations. As the *Washington Post*'s Catherine Rampell explained, many of the most notable changes to the new NAFTA were apparently "cribbed from another trade deal that Trump has demonized: the Trans-Pacific Partnership," to which Canada and Mexico were both party.[82]

The result was an awkward dynamic:

1. Trump condemned NAFTA, despite not knowing any relevant details about it.
2. Trump condemned the TPP, despite not knowing any relevant details about it.
3. Trump killed the TPP and demanded negotiations to revise NAFTA.
4. Trump settled on an agreement that, as Rampell put it, was "mostly just a smooshing together of two trade deals that he derided as the worst trade deals ever made."

The White House didn't even try to defend or explain any of this. The president had successfully slapped a new name on

NAFTA, and since Team Trump considers the substance of public policy an afterthought, little else mattered.

The American president then directed his attention to launching a new trade war, which was hatched in a characteristically irresponsible way. In March 2018 Trump got the ball rolling by blurting out plans to impose new tariffs on all steel and aluminum imports. NBC News ran a report soon after explaining that Trump made an impulsive decision "born out of anger" at unrelated issues, which had left him "unglued."[83]

Much of the president's economic team explained to him that tariffs on steel and aluminum would undermine the U.S. economy. His national security team told him the tariffs would run counter to U.S. national security interests. His diplomatic team concluded the policy would be bad for U.S. foreign policy. None of this stopped the nation's chief executive from risking adverse consequences for his own country because he was in a bad mood.

What's more, Trump blindsided members of his own White House team, who weren't informed about the tariffs in advance.

In a governing party, a major decision along these lines would be subjected to lengthy scrutiny and analysis long before a presidential declaration. In a post-policy party, no one at the State Department, the Treasury Department, or the Defense Department even knew that tariffs were coming. Conducting a meaningful review of the policy and its likely impact was out of the question. The president was winging it, despite his unfamiliarity with the subject matter, and there was nothing his aides could do about it.

The NBC News report added, "There were no prepared, approved remarks for the president to give at the planned meeting, there was no diplomatic strategy for how to alert foreign trade partners, there was no legislative strategy in place for informing Congress, and no agreed-upon communications plan."[84]

If a professor were teaching a class on how policy making

should never happen, she could do worse than pointing to the start of Trump's trade war.

But as little sense as the process made, the president's policies themselves were even more difficult to defend. At the launch of his agenda, Trump turned to Twitter to declare that trade wars are "good" and "easy to win"—a sentiment that suggested he wasn't altogether clear on what a trade war even was. Among trade economists, the idea that a series of escalating, tit-for-tat retaliatory measures between two countries is "good" for anyone is folly.

The president published a related tweet a month later, declaring, "When you're already $500 Billion DOWN, you can't lose!" Those who wondered how Trump managed to lose money running a casino suddenly had their answer.

All the while, it was entirely unclear what, exactly, Trump hoped his trade war would accomplish. Wars tend to have strategic purposes; the chief executive's trade agenda was incomprehensible.

The president seemed to operate from the assumption that a trade deficit with any country was both a problem in need of an immediate solution and evidence of a failed trade policy. "We lost, over the last number of years, eight hundred billion dollars a year," he said as his trade war got under way, shining a light on his fixation. "Not a half a million dollars, not twelve cents. We lost eight hundred billion a year on trade."[85]

It was a difficult assertion to take seriously. For one thing, trade deficits do not reflect "lost" money. As a *Washington Post* analysis explained,[86] in response to Trump's complaints about a trade deficit with China, "This is a strange sort of cost-benefit analysis that ignores the benefits of imports (what we get), and looks only at the costs of them (what we pay). It's like saying that we lose money anytime we go shopping anywhere, because stores aren't buying anything from us in return."

What's more, most economists simply don't see trade deficits as especially important. As a *New York Times* analysis put it,[87] "That's because trade imbalances are affected by a host of macroeconomic factors, including the relative growth rates of countries, the value of their currencies, and their saving and investment rates. For instance, America's trade deficit narrowed dramatically during the Great Recession, when national consumption faltered."

Even if Trump disagreed with all of this, there was little evidence that tariffs were helpful in narrowing a trade gap. By the time the president's third year in office was nearly complete, the overall U.S. trade deficit had grown considerably larger, not smaller.[88]

Adopting an exceedingly charitable tone, Zeeshan Aleem wrote for the news website Vox in March 2018, "The president doesn't seem to have a full grasp of what's going on here."[89] The assessment was more than fair: Trump found it difficult to explain why he was so concerned about trade deficits, whether he understood why trade deficits exist, and how his agenda would work in addressing a problem that most experts don't recognize as a problem. The Republican had ample time to get up to speed on what was supposed to be one of his strongest subjects, but he generally appeared to be unaware of the most basic details of the trade debate.

Trump surrounded himself with aides who weren't well positioned to help. The president's principal adviser on trade policy was economist Peter Navarro, who described his White House role to Bloomberg News this way: "My function, really, as an economist is to try to provide the underlying analytics that confirm [Trump's] intuition."[90]

As a rule, presidential advisers are expected to offer guidance based on research, judgment, and subject-matter expertise. An economist, in particular, is not supposed to start with someone's intuition and then work backward to bolster preconceived ideas. It's a classic example of post-policy thinking.

But in Trump's White House, this was the norm. When shaping life-changing policy, Trump and his team could rely principally on evidence, data, facts, and reason, or they could go with what strikes them as intuitively true. The president preferred the latter, making a clear distinction between technocratic governing and listening to the gut of a television personality turned amateur president with no background in government, public policy, or public service.

As for Navarro, *Vanity Fair* magazine reported in April 2017 on how the relatively obscure economist entered Trump's orbit: "At one point during the campaign, when Trump wanted to speak more substantively about China, he gave [his son-in-law Jared] Kushner a summary of his views and then asked him to do some research. Kushner simply went on Amazon, where he was struck by the title of one book, *Death by China: Confronting the Dragon—a Global Call to Action*, coauthored by Peter Navarro. He cold-called Navarro, a well-known trade deficit hawk, who agreed to join the team as an economic adviser."[91]

The result was far from encouraging. The *Los Angeles Times*'s Michael Hiltzik explained in April 2018 that Trump "hasn't articulated a coherent trade policy, and to the extent he has mentioned any principles, the steps he has taken won't serve them." He quoted Nicholas Lardy, a China trade expert at the Peterson Institute for International Economics, who commented, "The question that can't be answered is, what does the Trump administration want?"[92]

As the White House escalated trade tensions with China, it was a question that officials in Beijing asked frequently. Politico reported in June 2018 that Chinese officials were "increasingly mystified" by Trump's ambiguous trade goals.[93] A minister at the Chinese embassy in Washington said in a speech a week earlier, "We appeal our American interlocutors to be credible and consistent."

When two countries are in a trade war, and one of them hasn't the foggiest idea what the other is fighting for, there's a problem. Or as Paul Krugman put it in July 2018, "Is there a strategy here? It's hard to see one. There's certainly no hint that the tariffs were designed to pressure China into accepting U.S. demands, since nobody can even figure out what, exactly, Trump wants from China in the first place."[94]

The Atlantic's Annie Lowrey added six months later that Chinese trade negotiators were still "confused" by the Trump administration and its demands. "American officials raise issues only to later drop them," she reported. "They contradict one another. The ideological warfare within the White House, as well as the lack of experience on the international economic team, has left China and others unsure of U.S. policy, or even its goals."[95]

The embarrassment reached new depths in February 2019, when Trump hosted an on-camera Oval Office conversation with U.S. Trade Representative Robert Lighthizer and a delegation from Beijing, including Chinese vice premier Liu He. As the conversation progressed, Trump and Lighthizer ended up disagreeing with each other over the significance of formal "memos of understanding," leaving the Chinese leader to laugh audibly over the disagreement between the Americans—in the foreign delegation's presence—over the meaning of their own terminology.

All the while, the president had spent months assuring the public that China was pumping billions of dollars a month into the U.S. economy thanks to his trade tariffs. At times, Trump repeated the claim on a nearly daily basis, but the repetition didn't make the claim any less absurd: U.S. importers paid Trump's tax, and the costs were passed down to American consumers.

Even after White House officials acknowledged these details, Trump refused to believe them. Reality was counterintuitive to him, and so the president chose to ignore the evidence. He instead bragged about all of the things he'd do with the money from China that did not exist, vowing to apply the funds to everything from federal bailouts for farmers to the health care system.

In May 2019 the news website Axios asked several current and former administration officials whether Trump actually believed the falsehood about tariff revenue from Beijing.[96] The "consensus" was that the president had been entirely sincere, touting the nonsensical claim because he chose to believe it. One former aide told Axios that trying to encourage Trump to accept reality was pointless because the president's bogus beliefs were "like theology."[97]

The more transparent his confusion, the more certain Trump was that his expertise was unrivaled. In July 2018, referring to trade policy, he boasted, "I understand that issue better than anybody."[98] Four months later, he added, "I know every ingredient. I know every stat. I know it better than anybody knows it."[99]

It was around this same time that a *Washington Post* analysis was blunt in its breakdown of the president's incompetence and ignorance of the subject: "On trade, Trump either doesn't understand the basic facts or he doesn't care."[100]

Alas, that summarized the chief executive's post-policy approach to practically every substantive aspect of his presidency.

CHAPTER 3

"Even If It Worked, I Would Oppose It"

Health Care

IN HIS PRESIDENCY'S THIRD month, Barack Obama hosted a meeting with congressional Republicans at the White House, and the Democrat declared he was serious about pursuing an ambitious health care reform plan. GOP lawmakers, reminding the president about the rhetoric he'd used on the campaign trail a year earlier, demanded to know if Obama would follow through on his promises on bipartisan solutions.

The president responded by immediately putting a concession on the table: Obama didn't want to include limits on medical malpractice awards in the legislation, and he knew such a move would be unpopular with his party and some of its supporters, but in the interest of bipartisan cooperation and constructive legislating, the president told Republicans he'd agree to the concession if GOP officials were willing to make reciprocal compromises.

What, Obama asked, were Republicans prepared to offer? Nothing, GOP leaders responded. As *Time* reported in May 2009,

"Republicans were unprepared to make any concessions, if they had any to make."[1]

By any fair measure, the multiyear debate over health care was the Super Bowl of policy fights. Officials from both parties had spent the better part of a century pursuing measures intended to bring health coverage to those who lacked it, while improving health security for those with insurance, and policy makers had volumes of research and scholarship upon which to draw.

It was a once-in-a-generation opportunity, and the expectation was that Democrats and Republicans, each with firm ideas about the future of American health care, would bring their A games to the debate.

GOP lawmakers, however, wasted little time proving that they had no A game to bring. In what was arguably the defining domestic policy fight of the generation, Republicans were wholly ill-equipped for the debate, uninterested in engaging on any of the substantive details, and indifferent to the real-world consequences of their apathy.

This was hardly an abstract debate for tens of millions of American families. On the contrary, unlike most domestic political disputes, the fight over health care reform was a life-or-death issue. From those with preexisting health conditions to those who were one serious illness away from financial ruin, practically everyone would be affected by the outcome, creating ample incentives for GOP officials to be fully engaged. They chose a less constructive path.

For all the bluster from the right about Democrats having "rammed through" the Affordable Care Act as part of a "rushed" process, the health care reform law was among the most carefully scrutinized pieces of legislation in recent American history. The process featured dozens of congressional hearings across five committees as part of a bicameral debate that lasted more than a year.[2] In the U.S. Senate, after the committee debates, Democratic leaders held a floor debate that went on for twenty-five days.

As part of the months-long effort, Democratic officials spent countless hours meeting with stakeholders in the health care industry, holding town hall forums, and answering a myriad of questions from the public and the press.

The Republican approach was far less laudatory. For example, the idea of an individual mandate—a measure that requires the public to purchase insurance in order to control costs and spread risk—had been a staple of Republican policy making for decades. President Richard Nixon touted the idea in the 1970s, President George H. W. Bush supported it in the 1980s, and Bob Dole, the GOP's former Senate leader and former presidential nominee, endorsed it in the 1990s. In 2006 Mitt Romney, then the governor of Massachusetts, incorporated the individual mandate into this reform model in the Bay State, and he continued to promote the idea during his 2008 presidential campaign, the first of his two tries for the White House.

In 1993 Republican senator John Chafee of Rhode Island wrote a plan with an individual mandate that enjoyed the support of twenty other GOP senators, including Utah's Orrin Hatch and Iowa's Chuck Grassley.

In June 2009, with the Democratic reform initiative under way, Grassley went so far as to tell Fox News, "I believe that there is a bipartisan consensus to have an individual mandate."[3] He said two months later that the proper way to get to universal coverage was "through an individual mandate." The longtime Iowa lawmaker added, "That's individual responsibility, and even Republicans believe in individual responsibility."

It was after Democrats agreed to include the policy in their proposal that Republicans decided the individual mandate was the single most outrageous provision of the ACA and an unforgivable assault on Americans' liberties. Hatch, after having cosponsored a bill with an individual mandate years earlier, condemned the idea as "totalitarianism."[4]

When CNN asked the Utah senator in March 2010 to explain

the contradiction, Hatch replied, "Well, in 1993, we were trying to kill Hillary-care." He added that he "didn't pay any attention" to the part of the bill that included the mandate provision.

The cynicism surprised even jaded political observers, but Hatch's antics were also emblematic of his party's post-policy posture: the merits of proposals and the virtues of consistency were casually cast aside in the name of political expediency. If embracing the individual mandate meant trouble for Bill Clinton's reform initiative, great. If rejecting the individual mandate helped undermine Barack Obama's reform efforts, that was fine, too.

Grassley followed Hatch's lead. Four months after he told a national television audience about the "bipartisan consensus" on the individual mandate, and two months after he sang the policy's praises as part of a push toward "individual responsibility," Grassley was asked what kind of changes he'd like to see to the Democratic bill before he considered supporting it. "I'm very reluctant to go along with an individual mandate," he replied.[5]

There was no governing vision or genuine interest in addressing a life-or-death dilemma for millions of American families. This wasn't even a negotiation position from which to leverage related substantive gains. There was only a pathological desire to defeat a Democratic priority.

When partisans go to war against their own ideas after their rivals agree with them, the prospects for constructive bipartisan policy making disappear.

What's more, as the Affordable Care Act was taking shape, and Democrats pleaded with GOP lawmakers to partner with them on a bill, Republicans too often negotiated in bad faith. In the summer of 2009, the White House agreed to a significant delay in the reform process in order to allow the Gang of Six from the Senate Finance Committee to work on a possible compromise package. The group featured three Democrats (Montana's Max Baucus, New Mexico's Jeff Bingaman, and North Dakota's Kent

Conrad) and three Republicans (Grassley, Wyoming's Mike Enzi, and Maine's Olympia Snowe.)

The efforts proved pointless. Enzi hosted a discussion with a group of constituents in August 2009 and described his approach to the negotiations: "It's not where I get them to compromise, it's what I get them to leave out." The Wyoming Republican made it clear that he had no intention of supporting a reform bill and participated in the talks solely to weaken the bill he'd inevitably vote against.

When other GOP lawmakers tried to critique Obamacare and highlight its flaws, many Republicans seemed wholly unaware of the legislative details of the plan they purportedly hated. In September 2009 the National Republican Senatorial Committee sent out a letter to supporters over Texas senator John Cornyn's signature, claiming the ACA could establish a "lottery" system determining who would get priority treatment, while creating standards that allowed the government to discriminate against patients "on the basis of race or age."[6] The letter added that the idea of allowing federal officials to choose patients' doctors for them was "up for debate." Each of the claims were absurd, but the party made them anyway.

That same month, Representative Joe Wilson of South Carolina heckled Obama during a nationally televised address to a joint session of Congress—loudly shouting, "You lie!" at the president—because the right-wing lawmaker was convinced of the false idea that the health care reform law would cover benefits for undocumented immigrants.[7] Senator John McCain said in a written statement that the reform bill would add "more than a trillion dollars to our country's deficit," would put medical decisions "in the hands of government bureaucrats," and amount to a "government takeover of our health care system."[8] None of these claims had any relationship to reality.

House Minority Leader John Boehner inexplicably claimed

that the Democratic plan required "a monthly abortion fee."[9] Senator Rand Paul, an ophthalmologist before getting elected, argued that those who believe Americans have a right to health care were endorsing a more dangerous position: "That means you have a right to come to my house and conscript me. It means you believe in slavery. It means that you're going to enslave not only me, but the janitor at my hospital, the person who cleans my office, the assistants who work in my office, the nurses."[10]

Some arguments were little more than lazy examples of post-policy thinking: Louisiana Senator David Vitter said he had "a fundamental problem" with legislation that had a lot pages.[11] His GOP colleague Representative Roy Blunt of Missouri made the same case, as if the physical size of a proposal was relevant to its merits.

Fair and objective analyses made clear that there were real flaws in the Affordable Care Act. No law is perfect, and the architects of Obamacare had to make several difficult trade-offs, as is always necessary on legislation of this scope. But the plan's GOP critics largely ignored those substantive flaws, focusing instead on nonsense, much of which had been made up out of whole cloth.

In November 2009 the *Washington Post*'s Ruth Marcus marveled at the "appalling amount of misinformation being peddled" by Republicans. Her column added, "You have to wonder: Are the Republican arguments against the bill so weak that they have to resort to these misrepresentations and distortions?"[12]

The question soon answered itself, as GOP officials, unwilling or unable to participate in a real debate, launched a concerted effort to convince the public that Obamacare featured what it called "death panels."

At issue was an unremarkable provision in the bill intended to help elderly patients receive advice from their doctors on end-of-life directives and living wills. In July 2009 John Boehner claimed the provision "may start us down a treacherous path toward government-encouraged euthanasia."[13] The same week, Representative Virginia Foxx of North Carolina delivered a speech on the House floor, insisting that health care reform would "put seniors in a position" in which they may be "put to death by their government."[14]

This evolved into the death-panel talking point—the most disgusting of all the anti-ACA lies—touted by Alaska governor Sarah Palin, conservative media, and a few too many Republican members of Congress who knew better.

Senator Johnny Isakson tried to argue that end-of-life directives and living wills actually empower people "to be able to make decisions at a difficult time rather than having the government making them for you."[15] The conservative Georgia Republican added, "I don't know how that got so mixed up."

It was around this time that Chuck Grassley hosted a town-hall event in Iowa and was asked about the death-panel issue. "You have every right to fear," the senator told his constituents, adding, "We should not have a government program that determines if you're going to pull the plug on Grandma."[16] Asked on CBS's *Face the Nation* why he'd peddled such garbage, Grassley replied, "I was responding to a question at my town meetings. I let my constituents set the agenda."[17]

There was no great mystery as to how, in Isakson's words, things "got so mixed up." His GOP allies, unprepared for a proper policy debate, searched in desperation for something they could use to terrify the public and undermine support for the legislation.

It didn't work: the ACA was signed into law in March 2010. It wasn't until the dust had fully settled that the scope of the GOP's defeat came into sharper focus. Republicans had spent more than a year working from the assumption that if they stood firm, slapped away every outstretched hand offering an olive branch, opposed the ideas they used to support, lied uncontrollably, and refused to work in good faith, it would be enough to deny Obama and congressional Democrats a historic victory. They were mistaken.

Had GOP officials acted more like members of a governing

party, they would have been in a position to move the reform law in more conservative directions and include more of the party's goals—which Democrats would have gladly embraced in exchange for bipartisan support. Republicans chose a more self-defeating strategy that left them empty-handed. They didn't even have a rival health care solution to point to as evidence of their alleged seriousness on the issue.

Indeed, throughout the lengthy debate, Republicans would occasionally confront an awkward question: If they hated the ACA so much and had superior ideas about how to bring health security to Americans, why didn't they present an alternative reform plan of their own?

In June 2009 House GOP lawmakers presented a brief outline of their health care priorities, though it was vague to the point of comedy.[18] The document shed no light on who would benefit from the Republican approach, how much it would cost, or how it would be financed. The GOP plan was missing a key ingredient: a plan.

The day of the outline's unveiling, Roy Blunt, a member of the House Republican leadership and the chairman of what he labeled the House GOP "Health Care Solutions Group," declared proudly, "I guarantee you we will provide you with a bill."[19] The same week, Eric Cantor told reporters that the official Republican version of Obamacare was just "weeks away."[20]

In the years that followed, Jeffrey Young of the news website Huffington Post (now HuffPost) got quite a bit of mileage out of a running joke, documenting each of the many instances in which GOP officials said their reform plan was poised for its big unveiling. The list of examples wasn't short—some even included specific, self-imposed deadlines—but every promise about the plan's imminent release was broken.

All the while, Republicans seemed unable not just to package their own ideas but also to engage in a meaningful debate on the Democratic proposal. Several weeks before the final congressional votes on the ACA, the White House hosted a health care summit with the bipartisan congressional leadership from both chambers, whom the president invited for what he'd hoped would be a substantive discussion.

GOP leaders had very little to say. Toward the end of the event, John Boehner told Obama, "I've been patient. I've listened to the debate that's gone on here. But why can't we agree on those insurance reforms that we've talked about? Why can't we come to an agreement on purchasing across state lines? And why can't we do something about the biggest cost driver, which is medical malpractice and the defensive medicine that doctors practice?"

The trouble was, the policy makers sitting around the table had just spent hours answering each of those questions in excruciating detail. The House Republican leader wasn't posing these questions in an opening statement, hoping to lay the groundwork for additional discussion; he was posing questions that everyone in the room had already answered.

Writing for the the *New Republic,* Jonathan Chait observed, in reference to the future House Speaker, "It's like he wasn't even there. Does he not understand what the other side is saying? Does he not care at all? It's not that he's provided an answer to Obama's arguments that I disagree with. He's just totally unable to acknowledge or engage at any level with the arguments presented. You're debating a brick wall."[21]

The headline on Chait's column read, "Why You Can't Discuss Health Care with the GOP."

As the legislative process concluded, House Republicans did manage to slap together a halfhearted joke, which they brought to the floor as an alternative, but it was a plan in name only: it largely ignored the uninsured, looked past those with preexisting conditions, and offered nothing for those worried about losing coverage when it's needed most.

After examining the implications of the GOP proposal, Matt Yglesias, writing for the progressive news website ThinkProgress, described the Republicans' model as one that created a system that "works better for people who don't need health care services, and much worse for people who actually are sick or who become sick in the future. It's basically a health un-insurance policy."[22]

The Congressional Budget Office concluded that the plan, if approved, would have left roughly fifty-two million Americans without access to basic medical care by 2019.[23] Though the party's outline, unveiled months earlier, suggested Republicans would support tax credits for working-class consumers, the plan didn't bother to include the idea in the bill presented on the House floor.

This GOP alternative was easily defeated and quickly discarded by its Republican authors, who went back to assuring the public that the party was hard at work finalizing the details of a *real* alternative to the Affordable Care Act. In April 2012 Republican congressman Fred Upton of Michigan, in his capacity as chairman of the House Energy and Commerce Committee, told reporters, "Our wheels are beginning to turn."[24]

They turned all right, but at a glacial pace. It would be five more years before the public saw the fruits of the GOP's labor—in the form of a regressive Trump-era bill that failed in a Republican-led Congress.

While the party faced considerable mockery over the missed deadlines and broken promises, what was often overlooked was the explanation for the GOP's failures: the Republican Party struggled to come up with a health care plan because of its inability to function as a responsible governing party.

Every credible and effective solution to the nation's health care challenges required some combination of regulating the private insurance market and investing in broader coverage for consumers. Republicans had ideological objections to both courses: they had no use for subject-matter experts who steered them in directions they didn't want to go, and they lacked the wherewithal to come up with work-around solutions.

In April 2014, after House Majority Whip Kevin McCarthy of California announced the latest in a series of delays in the unveiling of his party's plan, a congressional Republican health care aide told Talking Points Memo that the party's policy makers were sincere in their desire to write a proposal, but they found it nearly impossible to avoid crafting a blueprint that ended up looking like the Democratic reform plan they were desperate to destroy.[25]

"As far as repeal and replace goes, the problem with replace is that if you really want people to have these new benefits, it looks a hell of a lot like the Affordable Care Act," the aide said, adding, "To make something like that work, you have to move in the direction of the ACA."

Republicans refused to do that, but they also refused to let go of their anti-ACA crusade. The tension led frustrated GOP leaders to believe the smartest course of action was several dozen repeal votes in Congress, each of which they knew in advance would amount to nothing.

The exact number of repeal votes was the subject of some debate, though in 2017 *Newsweek* magazine counted *seventy* instances in which Republicans voted to repeal the reform law, in whole or in part, across both chambers.[26] (One of the later votes was held on Groundhog Day, which, in light of the 1993 Bill Murray movie, added a touch of unintentional thematic humor to the farce.)

"I don't care," Congressman Rich Nugent declared in July 2013 about the incessant repetition. "If we do it a hundred times, sooner or later we'll get it right."[27] The Floridian didn't specify what, exactly, he wanted to get right or what the effects would be if he succeeded.

It was an exercise in self-indulgent vanity and an abandonment of anything resembling responsible governing. GOP lawmakers knew full well that they couldn't repeal the health care law while a Democrat was in the White House, but they just kept going through the motions, finding it easier than working on public policy.

In May 2013 GOP representative Mick Mulvaney of South Carolina characterized the votes as rites of passage for new members. "The guys who've been up here the last year, we can go home and say, 'Listen, we voted thirty-six different times to repeal or replace Obamacare,'" he said at a conservative gathering, four years before joining Donald Trump's White House as director of the Office of Management and Budget (and later, moonlighting as the president's third chief of staff). "Tell me what the new guys are supposed to say?"[28]

Many Americans are familiar with the line about insanity being defined as doing the same thing over and over again and expecting different results. But the true insanity of the Republicans' repeal votes is that they didn't care about the results. They were pandering to the party's radicalized base while scratching a partisan itch, uninterested in doing real work, and indifferent to how their efforts would affect the public if implemented.

The inconsequential nature of the time-wasting exercise was deemed irrelevant. For a post-policy party, repeatedly tackling meaningless bills in a clumsy political ballet is a perfectly legitimate use of congressional power.

In one instance after the reform law was passed, some moderate Senate Democrats expressed a willingness to scrap the ACA's individual mandate—the policy Republicans loved before they hated it—and replace it with some related measures that could achieve similar results.[29] The centrist Democrats generally assumed that GOP lawmakers, eager to repeal elements of the reform law, would jump at the chance.

They did not. Republican leaders concluded that the Democratic moderates were ripe electoral targets, and if voters saw these senators striking bipartisan agreements with their GOP colleagues, it might give them a reelection boost. Politico reported, "Some Republicans are quietly warning colleagues not to work with vulnerable Democrats," even if they offered to work constructively on a priority the GOP said it cared about.[30]

Or put another way, Democrats were interested in moving the Affordable Care Act in a direction Republicans liked—eliminating an individual mandate that the GOP saw as a catastrophic and unprecedented assault on freedom—but Republicans weren't interested. The party was given a choice between prioritizing an election strategy or a substantive goal. It focused on the former, even if the latter would have advanced its purported interests.

On some rare occasions, Republicans tried to use meaningful data to bolster their case against the reform bill but stumbled because of their unfamiliarity with government reports. In early 2011 the Congressional Budget Office (CBO) concluded that the projected "reduction of labor" as a result of the ACA would be "a reduction of eight hundred thousand workers" by the year 2021. Republicans pounced, convinced that they'd finally found a silver bullet.[31] Paul Ryan called it "more bad news for American families." The National Republican Congressional Committee said the news proved that Obamacare would "cost eight hundred thousand jobs."

Either GOP leaders didn't bother to look up what "reduction of labor" meant, or they knew the truth and didn't care. Either way, what the CBO had found was that, thanks to the Affordable Care Act, some workers would voluntarily change their employment status, since they would no longer have to work in jobs they didn't want simply to maintain health coverage.

For Democrats, this was a feature of the reform law, not a bug. The point was to end "job lock": a policy dynamic that routinely prevented people from retiring, starting new businesses, or moving to part-time status because they could ill afford to lose their benefits.

The CBO's findings about the law pointed to reductions in the labor supply, not the number of actual jobs. Republicans were hysterical, not because they'd uncovered devastating information about the ACA, but because they appeared confused about basic information. In a fact-checking piece, the *Washington Post*'s Glenn Kessler wrote, "This is the kind of political gamesmanship that gives politics a bad name."[32]

For GOP officials, nothing would prove more devastating to the Affordable Care Act than evidence of it undermining domestic employment. Republicans had already labeled the law a jobs killer, and all they needed was some kind of evidence to substantiate the claim.

It never arrived. In 2014, the first full year for ACA implementation, job growth in the United States reached a fifteen-year high. The tally in 2015 was even stronger, and it was the best year for job growth in the twenty-first century. Despite Republican predictions, the first two years in which Obamacare was fully in effect were the best back-to-back years for American job creation since the economic boom in the 1990s.[33]

What's more, Dan Diamond, writing for *Forbes,* discovered that during the postcrash recovery, the U.S. private sector last lost jobs in February 2010. The Affordable Care Act became law in March 2010—and American businesses hadn't lost jobs since.

Indeed, to the right's great frustration, Obamacare had a nasty habit of succeeding—and exceeding expectations—even as it faced withering attacks from its opponents. It wasn't long before each of the Republicans' dire predictions, based entirely on political assumptions instead of policy analyses, were thoroughly discredited.

GOP officials had expressed with great certainty that Americans wouldn't want to buy coverage through the Affordable Care Act's exchange marketplaces; the system would never meet its enrollment goals; private insurers would steer clear of the ACA system; and the uninsured rate would remain stubbornly high.

When reality disproved each of the predictions, Republicans didn't take it well.

In early 2014, when millions of consumers purchased coverage through the ACA, Senator John Barrasso was so reluctant to believe his lying eyes that he ran to Fox News to accuse Obama administration officials of "cooking the books."[34] He had no proof, and the Wyoming Republican was soon proven wrong. Nevertheless, others in his party, including Senator Lindsey Graham and radio shock-jock Rush Limbaugh, soon echoed the conspiracy theory. When reality suggested that Obamacare was working, Republicans determined reality could no longer be trusted.

Some GOP leaders decided the law's success wasn't altogether important. The day after Barrasso touted his conspiracy theory, John Boehner used his Speaker's perch to declare, good news be damned, "The president's health care law continues to wreak havoc. . . . House Republicans will continue to work to repeal this law."[35]

In the film *Monty Python and the Holy Grail,* villagers decide they want to burn a suspected witch, and John Cleese's character offers proof of her evil ways: "She turned me into a newt." It's obvious that he is not, in fact, a newt, leading Cleese to say sheepishly, "I got better."

Villagers, enraged and ignorant, exclaim moments later: "Burn her anyway!"

The scene came to mind after seeing Boehner call for the demise of a health care law that was obviously working. Republicans had already decided they wanted to destroy the law, citing dubious evidence that turned out to be wrong. Presented with reality, they nevertheless shouted, "Destroy it anyway!"

Others were willing to acknowledge the ACA's progress, but they believed it came at too a high price. Representative Pete Sessions, the chairman of the House Rules Committee, went to the House floor in March 2015 to share numbers he'd come up with. "If you just do simple multiplication," the Texan said, "twelve million [consumers] into $108 billion, we are talking, literally every single [ACA] recipient would be costing this government more than five million per person for their insurance. It's staggering. . . . One hundred eight billion for twelve million people is immoral. It's unconscionable."[36]

A key element of the post-policy thesis is the Republican Party's rejection of arithmetic in a policy-making context, and Sessions offered an unusually helpful example. For one thing, $108 billion divided by twelve million is $9,000, not $5 million. The congressman's per-person estimate was off by only $4,991,000.

For another, Sessions's numerator and denominator were both wrong: CBO data at the time pointed to twenty-three million consumers, covered at a cost of $95 billion, for an average cost of $4,130—which could hardly be described as "unconscionable."

The larger point went far beyond an arithmetic-challenged congressman. Sessions's speech on the congressional floor stood as one of countless examples of how difficult it was for federal policy makers to have constructive discussions about health care policy. The parties couldn't have meaningful conversations because Republicans had manufactured an alternative reality they found more ideologically satisfying.

As Vox founder Ezra Klein put it, "If you're a conservative and you consume conservative media, you now live in a world . . . so different from the one that Democrats share with the CBO that no argument is really possible. Democrats say the bill reduces the deficit. Republicans say that the bill explodes the deficit. And when the scorekeeper tries to intervene, Republicans take aim at the scorekeeper. Real debate isn't possible under those circumstances."[37]

It drove home the implications of the GOP's transition into a post-policy party: most Americans have some expectations that the nation's two major parties will at least try to work cooperatively and with a sense of shared purpose, but the more Republicans abandon the pretense that the substance of public policy matters, the more the window shuts on constructive, bipartisan governance.

In cases in which GOP officials didn't like official data, they pulled stunts to create their own. In 2014, after the White House boasted about the millions of consumers who enrolled in the ACA through exchange marketplaces, Republicans on the House Energy and Commerce Committee published a one-page report that sought to prove only a modest percentage of consumers who signed up for coverage had paid their first month's premium, which meant enrollment rates were artificially encouraging. The GOP-led committee said the figures were based on a questionnaire Republicans had sent directly to executives at private insurance companies.

The faux survey had been manipulated to a ridiculous degree in order to produce misleading numbers Republicans liked. One insurance industry official acknowledged, "The survey was so incredibly rigged to produce this result, it was a joke. Everyone who saw it knew exactly what the goal was."[38]

Mother Jones magazine's Kevin Drum said the scheme was "about as sophisticated as a kindergartner throwing a mud pie."[39]

It was widely assumed, at least at first, that congressional Republicans were simply trying to deceive the public with bogus data. What was truly amazing, however, was the fact that GOP lawmakers were so confused, and so uninterested in pretending to be a governing party, that the House Energy and Commerce Committee convened a May 2014 hearing with insurance company executives in order to bring attention to the GOP report that had already been discredited.

Democratic lawmakers could hardly believe their good fortune: insurers gave sworn testimony, at Republicans' behest, that cut Republicans' arguments off at the knees. They not only rejected the GOP report on the number of consumers who'd paid their first-month premiums, but also insurers proceeded to reject the idea that the Affordable Care Act represented a "government takeover" and to celebrate their companies' improved finances since the ACA had become law.[40]

Republican lawmakers were "visibly exasperated" by the testimony they'd arranged, wholly unaware of how badly it would backfire.[41] So why did the committee's majority party call the hearing in the first place? Apparently because its GOP members were so unaware of current events, so reluctant to brush up on the details before the hearing, and so convinced that the Obama administration and the liberal media were lying to them, that the Republican lawmakers didn't realize their propaganda wasn't true.

A year later, as the race for the Republican presidential nomination got under way, candidate and retired neurosurgeon Ben Carson, two years before becoming secretary of Housing and Urban Development in the Trump administration, not only compared the Affordable Care Act to slavery—perhaps he'd been chatting with Rand Paul—but also dismissed the relevance of the law's efficacy, saying the only principle that mattered was a private health care market free of public-sector interference.

"Even if it worked," Carson said of the ACA, "I would oppose it."[42]

The eight-word sentiment helped capture a fundamental difference between the parties. In any policy fight, Democrats tend to focus on policy outcomes and the material difference proposals make in people's lives. Republicans, as a post-policy party unconcerned with empiricism, disregard what's effective and what isn't.

As a consequence, when party leaders quarrel over an issue, they invariably talk past one another, not only offering different answers, but also asking different kinds of questions. On health care, the more Democrats spent a decade arguing, "The ACA works," the more Republicans responded, "It doesn't matter if it works."

As hapless as Ben Carson was on dealing with the substance of the debate, his principal rival for the GOP presidential nomination, Donald Trump, was considerably worse. In July 2015 the future president, who expressed hatred toward "Obamacare," though he struggled to explain why, assured the public that he'd replace the existing law "with something terrific." Asked to elaborate, the Republican said his system would depend on "private companies" that would excel through "competition."[43]

Or put another way, Trump used bluster to hide the fact that he had no meaningful understanding of health care reform. In September 2015 he appeared on CBS's *60 Minutes* and expressed great confidence in his ability to create a health care system in which "everybody" would be "taken care of much better than they're taken care of now." Asked how, Trump replied, "The government's gonna pay for it."[44]

By the time he'd locked up his party's nomination, the superficiality of Trump's approach had not improved. In April 2016 the *New York Times* reported that Trump's health care ideas had "bewildered" not just reform advocates but also conservative health care wonks: "This whipsaw of ideas [in Trump's plan] is exasperating Republican experts on health care, who call his proposals an incoherent mishmash." The article quoted Robert Laszewski, a former insurance executive and critic of the ACA, who characterized Trump's health care proposals as "a jumbled hodgepodge of old Republican ideas, randomly selected, that don't fit together."[45]

Trump and his team didn't seem to care and made no effort to defend their mess. It was a post-policy candidate, running on a post-policy platform, appealing to post-policy voters. It was also a harbinger of things to come.

During the presidential transition period, Trump and his allies continued to treat the debate in ways that suggested total indifference to substantive details, and the result was a series of instances in which Republicans wrote checks the party would never be able to cash.

In December 2016 Paul Ryan said the GOP policy—which did not yet exist—would ensure that "no one is left out in the cold" and "no one is worse off."[46] In a separate interview the same week, the House Speaker said the goal of his party's ACA repeal efforts was to ensure that "people can get better coverage at a better price."[47]

Parts of Ryan's pitch, however, descended into post-policy politics. Ahead of Inauguration Day, trying to help lay the groundwork for his party's eventual initiative, the Speaker told CBS, "We also think that a refundable tax credit is a smarter way to get people the ability to go buy insurance that they like that they can afford. That's better than [Obamacare's] subsidies."

In reality, the ACA's premium support subsidies were, in fact, refundable tax credits. As the online magazine Slate's Jordan Weissmann joked, Ryan's argument was the equivalent of a bartender telling a customer, "You'll like this drink way better, because I make it with vodka instead of Smirnoff."[48]

It was possible that Ryan was confused. Maybe he knew the difference and didn't care. Either way, comments like these suggested a degree of indifference to the governing details Republicans needed to understand to legislate properly on the issue.

Trump was similarly unburdened by concerns over policy details. In the days leading up to his inauguration, the president-elect boasted, "We're going to have insurance for everybody." He added that Americans "can expect to have great health care. . . . Much less expensive and much better."[49]

The president-elect even went so far as to establish specific benchmarks: universal coverage, "much lower deductibles," and a simpler and less expensive system in which all Americans are "beautifully covered."

Trump's ignorance about health care governance was briefly blissful. The Republican assumed that the principal problem with Obamacare was that it had been designed by fools who lacked what he perceived as his commonsense wisdom. Trump felt no constraints about making bold promises because, in his mind, he could take office, direct Congress to replace the ACA with a plan that offered better coverage at a lower price, lawmakers would follow his instructions, and the public would rejoice. As far as he was concerned, it was all quite simple.

Those who took governing seriously knew better. Policy makers have grappled for decades with systemic trade-offs, but the television-personality-turned-president assumed the challenges could be resolved to his satisfaction by the force of his will.

The president-elect's confusion about the legislative process didn't help. The week before Inauguration Day, as GOP congressional leaders considered a way forward on health care, Trump declared that he and his team would submit their own reform blueprint as soon as the Senate confirmed a new secretary of the Department of Health and Human Services. "We're going to be submitting, as soon as our secretary's approved, almost simultaneously, shortly thereafter, a plan," he said to the surprise of his allies on Capitol Hill, who had no idea the incoming White House was planning to submit anything of the kind.[50]

In fact, in January 2017, Republican leaders in the House and Senate coalesced around a new approach: a repeal-and-delay strategy in which the party would repeal the ACA quickly, and then give themselves a two- or three-year window in which to come up with an alternative policy model. The plan was the source of some embarrassment—it served as proof that GOP policy makers, after having started work on their alternative model in the summer of 2009, still had no idea what they were doing—made worse by the fact that it would have wreaked havoc on insurance markets, which rely heavily on predictable modeling.

"We want to make sure there is an orderly transition," Paul Ryan said, "so that the rug is not pulled out from under the families who are currently struggling under Obamacare while we bring relief."[51] The comment revealed a level of confusion the congressman was reluctant to address: if families were actually "struggling" under the ACA, then getting rid of the law wouldn't constitute pulling the rug from under them; it would mean the opposite. Ryan made no effort to reconcile the contradiction.

Soon after, the problem became a moot point. Without letting his own party leaders know in advance, Trump impulsively rejected the repeal-and-delay approach, derailing the Republican strategy without any real thought or deliberation.[52]

The president-elect added that he expected the GOP-led Congress to repeal the nation's existing health care law the week of his inauguration, with a replacement bill to follow "very quickly or simultaneously, very shortly thereafter." Trump was apparently unaware of the fact that his party didn't have a replacement bill, and the incoming White House team didn't have one to offer.

A month into his presidency, Trump aides alerted lawmakers to the fact that the administration had decided not to put together a health plan after all.[53] The president's earlier rhetoric on the subject was based on assumptions he didn't understand.

He wasn't the only one confused. On the heels of Republican leaders issuing assurances that "everybody" would have coverage and "no one" would be worse off, Republican congressman Mike Burgess of Texas, who chaired a key health care subcommittee, said at a conference that if his party's efforts increased the number of uninsured Americans, "I would say that's a good thing because we restored personal liberty in this country."[54]

It served as a reminder of the party's emphasis on ideological, not practical, considerations. In effect, some GOP lawmakers were prepared to tell the public, "Your family is one serious illness away from financial ruin, and your health is at risk from treatable ailments, but look at how great your liberty is."

As Republicans appeared lost without a map, the new president, five weeks into his term, reflected on the governing dynamic that left him baffled. "Very complicated issue," Trump said at a White House event with the nation's governors. "I have to tell you, it's an unbelievably complex subject. Nobody knew that health care could be so complicated."[55]

Generous observers argued that he was half right, since health care is inherently complex. Republicans had spent much of the decade attacking Obamacare while paying almost no attention to policy considerations, operating with the luxury of knowing that real work wasn't altogether necessary: they could vote on repeal bills, knowing they'd fail; and promise an alternative plan, knowing they could just push back their own deadlines. Their indifference carried no consequences.

The lack of real responsibilities was freeing—right up until they became the dogs who caught up to the car they were chasing. It was at that point that Trump and his party came to realize that the ACA wasn't just an appendage attached to the nation's health care system that could be removed with a legislative cleaver; it was interspersed by design throughout the system. Getting rid of it without hurting families, the economy, hospitals, states, insurers, and countless businesses would take effort, patience, policy expertise, and legislative dexterity.

Post-policy parties have no such skills. Republicans were trying to use muscles that had long since atrophied.

But while the new president may have been right about the intricacy of health care policy making, he was breathtakingly wrong about "nobody" being aware of this dynamic prior to February 2017. *Everyone* knew that health care policy was complex. It was complicated when Democrats spent more than a year crafting the Affordable Care Act and even more years trying to implement the reform law effectively. It was complicated when Republicans spent seven years trying and failing to write a credible alternative. It was complicated for generations, as policy makers in both parties launched various efforts to extend health security to Americans, starting in earnest in 1935, during Franklin Delano Roosevelt's first term.

Trump's capacity for surprise about the complexity of health care policy making betrayed a striking ignorance about a debate that had been ongoing for most of his life. It appeared the only person in the country who assumed health care policy was simple was the one in the Oval Office—following a campaign in which he ran on a platform of easily passing a great plan that would do more for less.

In late February 2017, House Republican leaders, at long last, took a step they hadn't taken before: Paul Ryan and his team circulated an actual piece of health care legislation that would have repealed the Affordable Care Act and replaced it with a regressive alternative. Though GOP leaders were serious about their proposal, it was difficult to take the bill itself seriously: Republicans intended to cut insurance subsidies for consumers, cut Medicaid spending that benefited low-income Americans, and scrap taxes on the wealthy that helped finance the existing ACA system. For good measure, the party's proposal also picked a culture war fight, defunding Planned Parenthood and eyeing measures that would have made it far more difficult for Americans to have abortions covered by private insurance.

The GOP plan not only failed to honor the lofty promises party leaders made a month earlier, but also it failed to try. The idea that the public would see the model as superior to Obamacare was plainly laughable.

Within days, Republican leaders killed the draft proposal and got to work on plan B. The party's American Health Care Act, nearly a decade in the making, was unveiled on March 6, 2017.

It landed with a thud. Avik Roy, a Republican health care adviser who has worked with several prominent GOP officials and candidates, could hardly believe his eyes. "Expanding subsidies for high earners and cutting health coverage off from the working poor: it sounds like a left-wing caricature of mustache-twirling, top-hatted Republican fat cats," Roy wrote in *Forbes*.[56]

In this case, however, the caricature was an accurate description of the GOP proposal. Structurally, and superficially, it resembled the Affordable Care Act, as both provided consumers with subsidies to buy private insurance. But the Republicans' alternative punished working-class Americans, rewarded the wealthy, and stripped the public of valuable consumer protections.

In a curious twist, an analysis from the Center on Budget and Policy Priorities, a progressive think tank, noted which states' residents stood to lose the most under the GOP plan.[57] Of the ten states that would fare the worst, all were red states that had supported Trump in 2016. CNBC's John Harwood noted, "'To the victor belong the spoils,' candidate Donald Trump reminded 2016 voters. But the House Republican health bill turns that old maxim on its head."[58]

On the surface, some core truths were obvious: Republicans wanted to repeal and replace the Affordable Care Act; they had to come up with an alternative solution; and they'd produced a highly controversial bill to further those goals. What was less obvious was the rationale behind this specific package.

Reason magazine's Peter Suderman wrote, "It's not clear what problems this particular bill would actually solve."[59]

Policy making at its most elemental involves identifying public challenges, weighing alternative solutions, and working toward effective results. By this definition, in early 2017, with Republicans controlling all of the levers of federal power, the GOP majority was barely even trying to govern responsibly. Vox's Ezra Klein added, "The biggest problem this bill has is that it's not clear why it exists. What does it make better? What is it even trying to achieve?"[60]

Those weren't necessarily rhetorical questions. Republicans had complained that the deductibles under the ACA were too high—but their plan would have pushed them higher. Republicans had spent months arguing that Obamacare didn't cover enough people—but their plan would have *increased* the number of uninsured. Trump committed publicly to a reform plan in which all Americans would be "beautifully covered"—but the GOP legislation did nothing of the kind.[61]

As the factions prepared for a spirted fight, post-policy politics emerged almost immediately. The day after the plan's unveiling, White House Press Secretary Sean Spicer stood alongside a table with two piles of paper, one significantly taller than the other.

The president's chief spokesperson wasted no time in delivering the punch line to an unnecessary joke: the tall pile of paper was the text of the Affordable Care Act, while the short pile was the Republicans' new American Health Care Act.

"People who have concerns about this, especially on the right, look at the size," Spicer said. "This [hand on tall pile of paper] is the Democrats'; this [hand on short pile of paper] is us. There is, you can't get any clearer, in terms of this [hand on tall pile of paper] is government; this [hand on short pile of paper] is not."[62]

He appeared to be quite serious. Just as some congressional Republicans had done in 2009, the White House press secretary argued with great sincerity that shorter legislation, regardless of its practical effects on people, is better—because it's shorter. Spicer declared, with a straight face, that the merits of national health care plans can be measured, at least in part, by page numbers. This was the state of the policy debate in the world's dominant superpower in 2017.

At the other end of Pennsylvania Avenue, in the nation's Capitol, the debate was no better. In fact, congressional Republicans didn't see the point in even having any debate at all: GOP leaders decided to advance the health care reform legislation without legislative hearings.

Republicans were so disinterested in scrutinizing their own plan and coming to terms with its real-world impact that they ruled out committee debates. There would be no testimony from subject-matter experts. Lawmakers would not have any opportunity to engage in discussions with industry stakeholders.

The more that industry leaders, including the American Medical Association, the American Nurses Association, AARP, and the American Hospital Association, denounced the Republican plan, the more party officials dismissed their perspectives as unimportant.

When Democrats worked on the ACA, the House members participated in 79 bipartisan hearings, heard from 181 witnesses, and accepted 121 amendments.[63] While considering the GOP's alternative, Republicans, who'd complained bitterly about Obamacare having been "rammed through," scheduled zero hearings, invited zero witnesses, and accepted zero amendments.

None of this alarmed the Republican-led White House or most of the rank-and-file members of the GOP conference. Representative Leonard Lance appeared on MSNBC's *All In* a week into the legislative process in 2017. When host Chris Hayes asked, "How many hearings—open hearings with witness testimony and the like—have you had on this bill?"[64] the New Jersey Republican hemmed and hawed, making every effort to avoid answering directly, reluctant to acknowledge that the correct answer was zero.

Finally, in response to repeated questioning, Lance said he expected the Republican-led Senate to engage in some "discussion" about the implications of the reform plan—*after* the House passed its bill without any meaningful scrutiny.

In the abstract, the purpose of a congressional hearing is to provide policy makers with information needed to improve legislation. Members of Congress, even those who've made up their minds, participate in hearings to explore issues in detail, hear from subject-matter experts, ask questions of authorities, dig into substantive nuances, and use the information gleaned from hearings to craft the best possible proposals.

When it came to overhauling the American health care system, Republicans saw this as an unnecessary step.

The policy debate, to the extent that one existed, was so hopeless that Paul Ryan seemed confused about the purpose of health insurance. "The whole idea of Obamacare is . . . the people who are healthy pay for the people who are sick," the Speaker argued. "It's not working, and that's why it's in a death spiral."[65]

The ACA was not in a death spiral—which occurs when healthy people exit the market, priced out by sicker consumers, who cost more to cover—and the congressman's comments suggested he was using the phrase without fully understanding its meaning.

But more importantly, by complaining that the ACA was based on the idea that the healthy pay for sick people's care, Ryan seemed to denounce the whole idea behind insurance itself.

Americans pay premiums, the money goes into a pool; funds from that pool pay for care. Those who are healthy don't draw from the pool—they instead pay for the sick—but the healthy do it anyway because no one ever knows when he or she might not remain healthy. Auto insurance works the same way: those who don't get into accidents pay for those who do, comfortable in the knowledge that if they have trouble, they'll be covered, too.

Ryan, in his zeal to attack Obamacare, also felt compelled to denounce Insurance 101—a curious move for the architect of his party's health insurance legislation.

It quickly became painfully obvious that Republican officials, facing a test of their own making, lacked the technical competence to craft, scrutinize, pass, and implement a health care plan they'd worked on for eight years. The week before the scheduled House vote, the nonpartisan Congressional Budget Office issued a report that found the Republicans' plan would add fourteen million Americans to the ranks of the uninsured within one year, and that number would expand to twenty-four million by 2026. Furthermore, the GOP legislation would end Medicaid as it existed.[66]

A governing party would have taken one look at the CBO's findings, hit the brakes, and recognized it as a back-to-the-drawing-board moment. Republicans shrugged their collective shoulders, went after the agency's credibility, and pressed on, ignoring the evidence.

They failed anyway. In March 2017 Paul Ryan pulled his legislation from the floor, confronted with near-universal opposition.

At the White House, aides argued that Donald Trump had tried his best to help. Sean Spicer told reporters the president "used every minute of every day" to get the reform bill passed.[67] In reality, the day before the vote, as GOP leaders scrambled to twist arms, Trump was on the South Lawn, having fun pretending to drive a large truck.

Even if Trump had invested the time in working the phones, rallying his partisan allies, and trying to persuade the public about the value of the plan, his efforts would have been in vain, in part because opposition to the bill was so fierce, and in part because the president was so disinterested in the substance of the debate that he knew practically nothing about the proposal.

At one point, Trump said of his party's plan, "It follows the guidelines I laid out in my congressional address: a plan that will lower costs, expand choices, increase competition, and ensure health care access for all Americans. This will be a plan where you can choose your doctor. This will be a plan where you can choose your plan."[68]

Not one of those claims was accurate. In fact, the descriptions were the opposite of the truth. The amateur president appeared to be describing a plan he wanted to see, which led him to pretend the proposal in his imagination was the one his party had put on the table.

Trump's cluelessness was such a problem that some White House officials were afraid to allow the president to engage in behind-the-scenes lobbying efforts, fearing he'd make important giveaways that would fundamentally change the legislation in ways he didn't understand.[69]

When Trump met with members of the House Freedom Caucus, a group of right-wing lawmakers with a reputation for demanding ideological purity, the members made clear that they saw their party's conservative bill as too liberal, because it still provided some subsidies to the uninsured. One lawmaker later told Politico that the caucus members realized from their Oval Office conversation that the president "didn't have sufficient command of the policy details to negotiate."[70]

Soon after, GOP leaders invited Trump to Capitol Hill to

deliver a pep talk, hoping it might help close the deal with on-the-fence lawmakers ahead of the floor showdown. The president told Republicans they had to pass the bill, not because it would help people, but in order to save face and honor amorphous "commitments" they'd made on the campaign trail months earlier. He couldn't encourage lawmakers to act on his policy preferences, because he didn't have any. He couldn't play a constructive role in the policy-making process, because he had no idea what that might entail. The president couldn't sell the bill's merits, because he wasn't aware of any.

After Trump spoke to the House Republican Conference, Representative Walter Jones of North Carolina left the closed-door meeting and said the president gave "no details" on policy.[71]

Another Republican lawmaker said members were "astonished" at Trump's inability to understand the substance of the debate in any meaningful way, adding, "He seems to neither get the politics nor the policy of this."[72]

It was a sentiment that has come up throughout Trump's presidency in reference to practically every issue.

Following the demise of the party's health care plans A and B, there was some talk of Republicans at least considering a more constructive approach to health care policy. "I don't want that to happen," Paul Ryan told CBS News in late March 2017.[73] The Speaker added that bipartisan policy making is "hardly a conservative thing."

The party's plan C, however, was most definitely a conservative thing.

A month after the American Health Care Act died, Republican congressman Tom MacArthur of New Jersey, who had a reputation as a relative centrist, quietly reached a deal with the House Freedom Caucus's chairman, Mark Meadows of North Carolina, to revive the discarded plan by moving the legislation even further to the right. The new iteration would allow states to sell plans that rolled back protections for Americans with preexisting conditions, while also making it easier for states to forgo the essential health benefits—covering areas such as preventive care, maternity care, and mental-health treatments—mandated by the ACA.

The "compromise" between a conservative GOP lawmaker and a not-quite-as-conservative GOP lawmaker appeared to be predicated on the idea that the party's far-right bill simply wasn't brutal enough.

It was still entirely unclear what problem Republicans were trying to solve or which segments of the population they were trying to assist. Worse, GOP officials were proceeding, once again, with a process that deliberately ignored those who knew what they were talking about.

In late April 2017 the *Los Angeles Times* reported that Republican policy makers had "almost completely shut out doctors, hospitals, patient advocates, and others who work in the health care system . . . despite pleas from many health care leaders to seek an alternative path that doesn't threaten protections for tens of millions of Americans."[74] The article quoted one insurance company official, who'd discussed reform efforts with GOP lawmakers, as saying, "They're not interested in how health policy actually works. It's incredibly frustrating."

Helping prove the insurer's point, Donald Trump expressed his delight over the revived repeal bill, though he still didn't know why. On CBS's *Face the Nation,* the president said the legislation featured a "guarantee" to protect those with preexisting conditions, describing a provision that did not exist.

A day later, Representative Dennis Ross, a Republican from Florida, said he'd spoken with House Majority Leader Kevin McCarthy about the party's legislative strategy, and the Californian told him, "It's no longer about what we should do or how we should do it. Now is the time to do it."[75]

The moment a majority party making life-or-death decisions concludes that the propriety of its choices no longer matters, it has forfeited its status as a governing party.

North Carolina congressman Robert Pittenger went a little further, making the case that his party's approach to health care policy making could disadvantage Americans with preexisting conditions, but he had a solution in mind: those people could be encouraged to move to more progressive states. "We're putting choices back in the hands of the states," Pittenger said.[76]

Bolstered by arguments like those, the Republican-led House managed to narrowly pass the GOP health care bill on May 4, 2017. Party leaders, in an ostentatious display of willful ignorance, brought the bill to the floor before receiving a score from the Congressional Budget Office, which meant lawmakers had approved legislation radically changing the American health care system without knowing what their plan would cost or how many people would lose coverage if it were implemented. The bill was rushed to the floor for a vote so quickly, members didn't have time to read it even if they'd wanted to.

After the vote, some Republicans pointed to the bill's passage as evidence that the party was capable of governing.[77] In fact, it had done the opposite: passing a cartoonishly pernicious piece of legislation in a ridiculous way, without the slightest regard for its merits. In the process, GOP officials inadvertently presented a case study in the breakdown of American governance.

Representative Chris Collins, a Republican from Buffalo, New York, admitted to CNN that he hadn't familiarized himself with the bill before voting for it. When the *Buffalo News* alerted him to a provision in his party's plan that would crush a program that served 635,000 people in his home state, Collins had no idea what the newspaper was referring to.[78]

Other GOP House members struggled in different ways. For example, after the vote, Mario Diaz-Balart of Florida said, "Is this bill good? No, I don't like it,"[79] while Michigan's Justin Amash conceded, "This is not the bill we promised the American people."[80] Both Diaz-Balart and Amash voted for it anyway, saying they saw the bill as a vehicle to move the process forward. (Two years later, Amash would leave the Republican fold and declare himself an independent due to his growing disenchantment with both the party and President Trump.)

That process brought the proposal to the Republican-led Senate, where the House bill had no realistic chance of success. Mitch McConnell, the chamber's GOP leader, promptly created a thirteen-member "working group"—made up entirely of conservative men—tasked with writing yet another reform package. In the weeks that followed, no evidence emerged of the group ever having done any real work.

Instead, Senate GOP leaders worked in secret, without a hint of outreach to industry stakeholders and, once again, without any legislative hearings.

According to Don Ritchie, the U.S. Senate's official historian for nearly four decades, Senate Democrats crafted major tariff reforms in secret during the administration of Woodrow Wilson (1913 to 1921). But, he added, "Not since the years before World War I has the Senate taken such a partisan, closed-door approach to major legislation."[81]

Some in the GOP expressed modest discomfort. Senator Bob Corker said of his party's process, "There are no experts. Typically, in a hearing, you'd have people coming in, and you'd also have the media opining about if a hearing took place, and X came in and made comments." The Tennessean added that a public process generally helps "shape policy."[82]

But that's true only when policy makers care about crafting worthwhile legislation in a responsible way. And despite Corker's concerns, he made no effort to improve the process or block his party's repeal plans.

His colleagues' apathy was even more acute. In June 2017, reporters from Vox asked eight Senate Republicans what problems in the health care system they believed their party's reform bill would address.[83] None of the eight was able to provide a straightforward answer. Some didn't seem to understand the point of the question.

The same week, Oklahoma senator James Inhofe said of his party's plan, "I'm not sure what it does. I just know it's better than Obamacare."[84] Asked how and why it'd be better, the Oklahoman ducked into an elevator.

In many ways, Trump was worse, because he'd convinced himself he could speak intelligently about the legislation, despite the evident fact that he was effectively illiterate on the subject. The *New York Times* spoke to one senator who left a White House meeting with the impression that the president "did not have a grasp of some basic elements of the Senate plan."[85] A *Weekly Standard* report added that "several" Senate Republicans found after speaking to Trump that he had "little apparent understanding of the basic principles of the reforms and virtually no understanding of the details."[86] The *Washington Post* added that "seasoned senators," after speaking with Trump, "saw a president unable to grasp policy details."[87]

One senator said the president's vocabulary on health care was limited to roughly "ten words."[88]

Despite his inability to engage in meaningful negotiations with members of his own party, the president nevertheless boasted, "You have to know your subject. . . . In a short period of time, I understood everything there was to know about health care."[89]

With the hapless president stuck on the sidelines, unable to even feign competence, McConnell's bill collapsed in mid-July. Unfazed, Senate GOP leaders moved forward a week later with plan D.

The amazing thing about the fourth congressional Republican repeal-and-replace bill was that no one knew what it was, only that the Senate was poised to vote on it. In late July 2017 Fox News anchor Chris Wallace asked Senator John Thune, "As the number three Republican in the Senate, do you know what you're going to be voting on next week?" The South Dakotan wouldn't answer directly, saying only that Mitch McConnell would figure out something "at some point."[90]

Just forty-eight hours later, the Senate approved a curious procedural motion: members agreed to proceed with a debate on yet another Republican health care plan, which didn't yet exist.

None of this seemed to bother most GOP senators—members of an institution known as the "world's most deliberative body"—some of whom said they didn't care what they were being asked to vote on. Senator Richard Burr of North Carolina, asked about being in the dark on the specifics of his own party's health care plan, said, "It doesn't concern me. As I said, I'll vote for anything"[91]

Despite Burr's willingness to prove the post-policy thesis true, McConnell's secret bill, called the Better Care Reconciliation Act, also failed. The GOP leader quickly shifted his attention to plan E, the Obamacare Repeal Reconciliation Act, and it failed, too.

Plan F was a less ambitious health care plan called the Health Care Freedom Act—a narrow bill that targeted the ACA's individual mandate—which came to be known as "skinny repeal." Republicans were unrestrained in their criticisms of the legislation, with Lindsey Graham, among others, calling it a "disaster" and a "fraud."[92] He and his GOP brethren were nevertheless prepared to pass it, so long as the House agreed not to do the same. These senators insisted their sole motivation was to keep the process moving and allow Republicans to try to figure out an entirely new model that looked nothing like the ones they'd just considered.

The final vote on skinny repeal was set for July 28, 2017. As of July 27, the bill hadn't been written, much less read or scrutinized. The strategy called for Republicans to write the bill over lunch, then pass it that evening, all with the expectation that the bill would never be implemented or signed into law.

The idea of carefully considering health care legislation, perhaps with a legislative hearing or two, was never entertained. In their near-fanatical drive to undermine Obamacare, Republican policy makers had seemingly lost their minds, adopting an approach to legislating unlike anything in the American tradition.

It was impossible to defend. Ross Douthat, a conservative columnist for the *New York Times,* expressed some sympathy for the GOP's underlying cause, but added, "To vote 'yes' under these circumstances—dead-of-night vote for a bill with no scrutiny that nobody wants—is just disgraceful."[93] Democratic senator Chris Murphy called the gambit "weapons-grade bonkers."[94]

Three Republican senators—John McCain, Susan Collins and Alaska's Lisa Murkowski—joined Senate Democrats in ending the madness and defeating the bill. Forty-nine GOP senators, representing 94 percent of the Senate Republican conference, were comfortable going along with their party's scheme, but they fell one vote short.

The latest failure compelled Mitch McConnell to declare, "It's time to move on."[95] Health care advocates could at last sigh in relief—until GOP lawmakers started the post-policy show anew two months later.

In September 2017 Senator Bill Cassidy of Louisiana partnered with Lindsey Graham on yet *another* repeal bill, which by some measures was the most radical to date. The previous iterations of Republican repeal-and-replace efforts were based loosely on the ACA chassis, but Graham-Cassidy intended to unravel the national model altogether and allow states to go in disparate directions.

In practical terms that meant scrapping the existing Medicaid program, ending federal protections for Americans with preexisting conditions, eliminating the ACA's requirements on essential medical benefits, and cutting federal subsidies to those covered through Obamacare.

Adding an additional wrinkle, for procedural reasons, if Republicans were going to pass this plan, they had to do so before the end of the fiscal year, which meant GOP senators had about a week and a half to write, read, scrutinize, debate, and pass an entirely new proposal that had the potential to hurt tens of millions of people.

Graham may have been the coauthor of the bill, but he generally demonstrated a Trump-like understanding of his plan. One senior Republican staffer on Capitol Hill conceded, "If there was an oral exam on the contents of the proposal, graded on a generous curve, only two Republicans could pass it. And one of them isn't Lindsey Graham."[96]

Chuck Grassley wasn't in a position to pass the exam, either, though he seemed determined to downplay the significance of the legislation's actual provisions. "You know, I could maybe give you ten reasons why this bill shouldn't be considered," the Iowan said. "But Republicans campaigned on this so often that you have a responsibility to carry out what you said in the campaign. That's pretty much as much of a reason as the substance of the bill."[97]

It was a perspective devoid of any governing value. When officials believe the substance of public policy is of equal significance to vague campaign rhetoric, they've lost the plot. Regardless, most of the GOP remained indifferent to potentially life-changing details. In the run-up to a vote, Axios reported, "Senate Republicans are on the verge of passing a sweeping health care bill not only without knowing what's in it, but without particularly caring." The article quoted a GOP lobbyist, who said, "I am just in shock how no one actually cares about the policy anymore."[98]

As the September 30 deadline neared, Republican senators didn't know what Graham-Cassidy entailed, how much it would cost, or how many Americans it would leave uninsured. They didn't know how to rebut arguments from doctors, nurses, hospitals, insurers, and patient advocates, each of whom fiercely condemned the legislation. They couldn't persuade the American mainstream on its merits. They couldn't hold a legitimate hearing debating its value.

Nevertheless, most Senate Republicans were prepared to pass Graham-Cassidy and impose it on the country—but not enough of them. On September 26, 2017, party leaders pulled the plug, unable to muster the necessary votes.

As the dust settled, Philip Klein, writing for the conservative *Washington Examiner,* concluded, "Don't look to the GOP for a rational and coherent alternative to expansive government, because Republicans offer nothing more than a health care strategy designed by idiots, full of sound and fury, achieving nothing."[99]

His palpable disappointment was rooted in an unavoidable fact: Klein seemed to expect Republican policy makers to approach the substance of health care governing with the same seriousness that he did. Unable to shake their post-policy approach to governance, they failed spectacularly to do so.

After his party exhausted its legislative options, Donald Trump and his team were left with limited choices, which were pursued with a familiar passivity toward responsible governance. The approach the president adopted with the greatest enthusiasm was a sabotage campaign against his own country's health care system.

GOP officials and their allies had experimented with deliberate attempts at undermining the ACA in 2013, with limited success. Some Republicans launched a public "Refuse to Enroll" campaign, for example, in which Americans were encouraged to go without health coverage, on purpose, as part of an ideological crusade against Obamacare.[100]

When rank-and-file conservative voters proved reluctant to take one for the team, some GOP lawmakers announced that their offices wouldn't help constituents who called seeking guidance on how to get coverage through the Affordable Care Act.[101]

After the Obama administration reached out to sports leagues to partner on a public-information campaign about the new law—similar to the way Governor Mitt Romney's administration partnered with the Boston Red Sox following the passage of Massachusetts's reform law in 2006—congressional Republican leaders contacted officials with the National Football League, Major League Baseball, the National Basketball Association, the National Hockey League, the Professional Golfers' Association, and NASCAR, demanding that they reject any overtures.[102]

In November 2013 Politico published a report highlighting what it described as the Republican Party's "calculated sabotage" initiative against the health care reform law.[103] The analysis added, "That may sound like a left-wing conspiracy theory . . . but there is a strong factual basis for such a charge."

The same week, an official in the Obama White House said, in reference to the administration's arduous efforts to get the ACA up and running, "You're basically trying to build a complicated building in a war zone, because the Republicans are lobbing bombs at us."[104]

At the time, GOP officials had limited authority, and the impact of their sabotage efforts were modest. Obamacare managed to thrive despite the Republicans' unprecedented campaign. But with the arrival of the Trump era, it was a very different story. In July 2018 the *New York Times* published a roundup of the most notable steps the Republican administration had taken to undermine the Affordable Care Act, and the list wasn't short.[105] From slashing funding for the "navigators" program to benefit those who needed help enrolling, to ending "risk adjustment" payments to insurers taking on potentially costly new customers, to hiding ACA benefit information on official government websites, Trump's efforts had all the subtlety of a bullhorn, and according to private insurance executives, the president's agenda pushed prices higher for consumers.[106]

Responsible governance, at its most basic level, includes faithful execution of the nation's laws. It was a standard the GOP administration chose to ignore.

In June 2019 Trump appeared on NBC's *Meet the Press* and made the case that the public should arguably see him as a savior of the health care law he'd fought so hard to extinguish.

"I could have managed Obamacare so it would have failed or I could have managed it the way we did so it's as good as it can be," the president said, adding, "I had a decision to make. I could have politically killed Obamacare. I decided not to do it."

The comments reinforced suspicions that Trump was confused about the policies unfolding around him. Two months into his presidency, he had declared the death of the Affordable Care Act, telling the public, "It's a dead health care plan. It's not even a health care plan. . . . Obamacare is not an alternative. It's not there. It's dead. It's dead."[107]

In the months that followed, Trump repeated the rhetoric obsessively, telling anyone who'd listen that the health care reform law is "dead." "Gone." "Absolutely dead." "Finished." A "dead carcass."[108] When few found the rhetoric compelling, the president chipped away at the ACA using executive powers he didn't fully understand.

But when Trump proved too incompetent and too incapable of effective governing to repeal and replace it, and Obamacare persevered despite his efforts, the president expected the public to believe he'd "decided" not to allow the Affordable Care Act to die—even as his administration pursued legal challenges that would gut the reform law.

Trump wasn't asked to explain how he arrived at that decision, which was just as well. He wouldn't have been able to say anyway.

CHAPTER 4

"Extending a Middle Finger to the World"

Climate Change and Energy Policy

AS 2018 CAME TO a close, the Office of the Director of National Intelligence prepared a summary on global threats, which featured terrifying assessments related to climate change. The report described climate hazards "such as extreme weather, higher temperatures, droughts, floods, wildfires, storms, sea level rise, soil degradation, and acidifying oceans," which are "intensifying, threatening infrastructure, health, and water and food security."[1]

The climate crisis, U.S. officials added, would contribute to global "economic distress and social discontent," while raising the prospect of environmental refugees. It was a stark reminder about what makes the issue critically important to the future of humanity.

About a decade earlier, during the 2008 presidential race, one of the major-party nominees was unapologetic about his "unequivocal" belief in global warming. The candidate insisted that the United States "needs . . . a cap-and-trade system"—a position he repeated throughout the campaign.[2]

Confronted by occasional climate deniers, this presidential hopeful had a smart answer at the ready: even if all the science was completely wrong, he'd say, addressing the climate crisis would lessen our dependence on foreign oil and create good-paying jobs in a burgeoning industry. And if the science was right, addressing the climate crisis would help save the planet, too.

The same candidate ran a televised campaign ad featuring wind turbines, solar panels, melting glaciers, and polluting smokestacks, while a narrator reminded viewers that the senator had stood up to George W. Bush and "sounded the alarm on global warming."[3]

The candidate's name was John McCain, the Republican Party's nominee in 2008.

Two years removed from his national defeat, the Arizona senator decided his sensible approach to the issue was in need of an overhaul. "I think it's an inexact science, and there has been more and more questioning about some of the conclusions that were reached concerning climate change," McCain said while campaigning for an ally in New Hampshire in October 2010.[4] The veteran GOP lawmaker added that unnamed "outside influences" could be responsible for "flawed" scientific conclusions.

A few years later, in January 2015, Democratic senator Brian Schatz of Hawaii pushed a symbolic measure that would have put the Senate on record agreeing that climate change "is real; and human activity significantly contributes to climate change." Republican senator James Inhofe, one of the nation's preeminent climate deniers, helped lead his party's opposition, arguing that the fight over climate science was a result of a conspiracy hatched by the United Nations. (Two weeks earlier, Senate Republican leaders had put Inhofe in charge of the committee overseeing environmental policy.)[5]

When it came time to vote on Schatz's measure, McCain, seven years after stating his "unequivocal" belief in global warning, voted with Inhofe. Fifty senators ended up supporting the amendment, but it failed in the face of a GOP filibuster.

Perhaps no issue better defines the Republican Party's indifference toward governing, evidence, and responsible policy making than its approach to climate science. The more GOP officials have confronted terrifying proof and pleas to act, the more they've dismissed the issue, burying their heads in the eroded soil.

When the global issue came to the fore in the 1990s, GOP policy makers were initially comfortable denying the evidence. In time, the party's position evolved, acknowledging the planetary temperature changes but clinging to the belief that human activities had nothing to do with the shifts. When that, too, proved unsustainable, some Republicans settled on the idea that the planet was warming, and people bore responsibility, but doing anything about the problem would be too difficult and costly, so there was no point in even trying.

As post-policy politics overwhelmed the party, this gradual evolution collapsed into something simpler, dumber, and more dangerous. As *Mother Jones*'s Kevin Drum wrote in a memorable 2012 analysis, "Conservatives have more recently backpedaled not just a single step in this process, but all the way back to the Paleolithic era." The right's new standard line on climate change, Drum added, was that global warming "is the biggest hoax ever put over on the American public."[6]

There was a period in which progress seemed possible. Toward the end of the Bush-Cheney era, the question in policy circles wasn't whether to address the climate crisis, but how. Some on the left were comfortable with proposals to simply impose strict limits on carbon emissions on the private sector, telling polluters they would have no choice but to comply. For some on the right, a cap-and-trade system—involving pollution limits and emissions permits that could be traded among plants and businesses—offered a more conservative, market-oriented solution to the same problem.

Indeed, it was a model with a Republican pedigree: the Reagan administration used a cap-and-trade system when phasing out consumer access to leaded gasoline, and the administration of George H. W. Bush and Dan Quayle relied on a cap-and-trade model when combating acid rain a decade later. For those in the GOP willing to accept the evidence, it stood to reason a similar approach could be equally effective when addressing carbon pollution.

Democratic policy makers slowly came around to the idea, seeing the market-based approach as a bipartisan compromise that would go a long way toward solving the problem. This led to a brief window of opportunity in which prominent Republicans, many of whom were seeking high offices, answered questions about global warming by endorsing cap-and-trade systems. Among presidential hopefuls, McCain's advocacy stood out, but in 2007 Mike Huckabee, the former governor of Arkansas, also endorsed the policy.[7] Another ex-governor, Tim Pawlenty of Minnesota, not only backed the idea publicly but also appeared in an Environmental Defense Fund commercial in support of a cap-and-trade plan, alongside yet another former governor, Democrat Janet Napolitano of Arizona.[8]

On Capitol Hill, Representative Mark Kirk, a Republican from Illinois, voted for a cap-and-trade plan in 2009 ahead of his successful 2010 U.S. Senate campaign. A year earlier, in advance of his own U.S. Senate race, Florida state lawmaker Marco Rubio endorsed giving his state's environmental agency a "mandate" to design a "cap-and-trade or carbon tax program."[9] In 2009 Representative Fred Upton of Michigan pushed for carbon-emission reductions and described climate change as "a serious problem that necessitates serious solutions."

But after the 2010 midterm elections, in which voters rewarded Republicans with control of the U.S. House, the window of opportunity slammed shut. Upton wanted to become chairman of the House Energy and Commerce Committee, so he rejected what he'd said months earlier, partnered with climate deniers, and denounced cap-and-trade measures.[10] Kirk contended that "ignorance" had led him to vote for such an idea, while Rubio became one of the Senate's staunchest opponents of climate action. Tim Pawlenty went so far as to say "it was stupid" for him to have endorsed a cap-and-trade proposal, and he expressed deep regret ahead of his bid for national office in 2012.[11]

When Republicans were prepared to be sensible about climate policy making, convinced that some actions were probably inevitable, they saw cap-and-trade as a solution that responded to the science and the scope of the problem, while honoring conservative principles and party orthodoxy. But as the GOP descended into post-policy status, Republicans drew a different kind of conclusion: maybe dealing with the crisis wasn't inevitable after all. If the party could agree en masse to reject the science, ignore climate change experts, tout conspiracy theories, embrace the positions of the polluters who contributed generously to their campaigns, and oppose any and all legislative proposals, then it wouldn't even have to bother with the conservative approach to tackling the crisis.

It was easier for GOP officials to simply stop caring, give up on governing on the issue, and turn the fight into a lazy culture war and proxy fight against caricatures of environmentalists. The Republican base, taking its cues from the party's elected officials, started looking at climate change as a partisan dispute, and ambitious GOP politicians soon learned that taking the issue seriously led to electoral trouble—because in party primaries, the GOP base looked askance at candidates who agreed with Democrats about what the right sees as an environmental hoax.

In the first year of Barack Obama's presidency, Democrats took up a cap-and-trade bill called the American Clean Energy and Security Act (ACES Act), which was based on many of the same principles Republicans had previously endorsed, including the McCain-Palin platform from the year before. That did not, however, prevent the legislative debate from descending into a train wreck.

The more the Democratic majority pointed to scientific evidence and official reports, the more GOP lawmakers responded with arguments wholly detached from public policy. For example, Representative John Shimkus asked during a committee hearing, "If we decrease the use of carbon dioxide, are we not taking away plant food from the atmosphere?"[12]

The same week, the Illinois Republican said there was no need for officials to address the crisis because supernatural powers could intervene to protect the planet. "Man will not destroy this Earth. This Earth will not be destroyed by a flood," he declared, adding, "God's word is infallible, unchanging, perfect."[13]

At a different congressional hearing, Texas Republican congressman Joe Barton tried to make the case that if global warming worsened, humanity could simply pursue an "utterly natural reflex response to nature," by finding "shade."[14]

Much of the congressional debate featured GOP lawmakers who seemed awfully confused about the proposal on the table. Todd Akin of Missouri was apparently under the impression that a cap-and-trade bill was an attempt to control seasons. "I mean, we just went from winter to spring," the far-right congressman said.[15] "In Missouri when we go from winter to spring, that's a good climate change. I don't want to stop that climate change, you know." With displays of judgment like this, it was a bit jarring when congressional Republican leaders put Akin on the House Committee on Science, Space, and Technology. He served on the panel alongside GOP colleague Dana Rohrabacher, a California congressman who claimed that climate science is "a total fraud" concocted by liberals who "want to create a global government."[16]

Another member of Missouri's delegation, Republican senator Kit Bond, suggested cap-and-trade proponents might be trying to eliminate carbon altogether. "Without carbon, my trees would die," he pointed out. "Carbon occurs naturally."[17]

House Minority Leader John Boehner, the year before becoming House Speaker, told a national television audience, "The idea that carbon dioxide is a carcinogen that is harmful to our environment is almost comical. Every time we exhale, we exhale carbon dioxide."[18] While it's obviously true that humans exhale carbon dioxide, the Republican seemed to have no meaningful understanding of the greenhouse gas effect or the global consequences of excessive carbon pollution.

It would have been problematic had the Ohioan intended to address the climate crisis with ineffective solutions, but with the congressional debate under way, Boehner seemed lost about the underlying questions themselves. The GOP leader thought the cap-and-trade debate was about treating carbon pollution as a carcinogen, or cancer-casing agent (it wasn't), and that carcinogens harm the environment. (They don't.) Boehner's office was nevertheless so pleased with his television appearance that his aides promoted it online. The fact that he didn't know what he was talking about wasn't seen as a problem.

A standard assumption at the time was that Republicans were merely playing dumb in order to defeat the cap-and-trade legislation. Their idiotic arguments were cynical and insincere, the argument went, and served as part of a political ploy launched by policy makers who knew full well that their talking points were ridiculous. The organized foolishness was simply a means to an end.

But as a practical matter, their sincerity or lack thereof was beside the point. In a critical legislative debate on one of the world's most pressing life-or-death issues, Republican members of Congress couldn't bring themselves to act like members of a governing party—even after Democrats embraced a proposal that many prominent GOP officials had previously endorsed.

As the climate crisis intensified, Democrats were committed

to finding a solution, while Republicans took up a post-policy posture that made a meaningful debate impossible. Even when the GOP tried to keep up appearances and point to an academic-based statistic that ostensibly helped bolster its passivity, the party flubbed it.

During the debate over the ACES Act, Republicans obsessively told the public that a cap-and-trade model would impose, on average, a $3,128 annual financial burden on the typical American home. The wildly misleading figure came from a wildly misleading interpretation of a study conducted by John Reilly, a scientist at the Massachusetts Institute of Technology.[19]

In March 2009 a House GOP aide reached out to Reilly directly, asking him to offer congressional testimony against the Democratic plan. The MIT scholar not only explained that he *supported* the cap-and-trade bill but also urged Republicans to stop spreading false claims based on his scholarly work, which party officials clearly didn't understand. GOP leaders ignored the appeal.

In June 2009 the Congressional Budget Office published a report on the expected costs of the Democratic plan and found that the average household could expect to pay about $165 per year as part of the effort to combat the climate crisis, not $3,128.[20] Lower-income families would pay less than nothing: according to the CBO, these households would actually get money back in the form of rebates through the cap-and-trade plan. None of this factored in the related financial benefits of an improved global climate.

The CBO's report added that as carbon permits were sold, and proceeds were returned to taxpayers, the overall costs would sink even lower over time.

The number touted by Republicans was off by a factor of 19. And yet, presented with the CBO's findings, GOP lawmakers shrugged and continued to push the bogus statistic.

The ACES Act narrowly passed the House but died in the Senate. Despite the intensifying planetary threat, it would be a decade before Congress even tried to vote on another major proposal intended to address global warming.

After losing badly in a 2010 congressional Republican primary, Representative Bob Inglis of South Carolina reflected on his party's willingness to consider evidence. "What we have been doing so far is sort of shrinking in science denial and holding on to shaky ideology that really will be overwhelmed by the facts," the former GOP lawmaker said. "You can hold back the facts only for so long, and eventually they overwhelm you. . . . I think that eventually the champions of free enterprise, which is who conservatives are, who Republicans generally are, will rise to the occasion and come forward with real solutions here."[21]

The world is still waiting.

Once Republicans controlled some of the levers of federal power, the prospects for meaningful legislation on climate change and energy policy collapsed, but GOP lawmakers weren't prepared to ignore the issues altogether. In July 2014, for example, Republicans were so eager to promote offshore oil drilling, despite the need to move toward cleaner and renewable sources of energy, that the GOP-led House approved a measure ordering the U.S. Department of Energy to stop blocking permits for offshore oil drilling.

What the authors of the measure failed to realize is that the Department of Energy doesn't regulate oil drilling at all.[22] It is the U.S. Department of the Interior that considers the permits. Representative Marcy Kaptur, a long-serving Democrat from Ohio, tried to explain to House members that the proposal didn't make sense, but its Republican supporters, unmoved by the details about their own government's agencies, couldn't be swayed. It made about as much sense as voting to require the Department of Education to start approving drilling permits, too.

The measure's Republican authors could have checked to see what the Department of Energy does before advancing the proposal to the floor. They also could have pulled the measure after being told about the differences between the Departments of Interior and Energy. But GOP House members proceeded, confident in the knowledge that their gut-level understanding of the federal bureaucracy was correct, even if reality said otherwise.

Related legislative efforts from that era were less amusing. Also in 2014, Republican congressman Jim Bridenstine of Oklahoma, years before Donald Trump put him in charge of NASA, championed what he called the Weather Forecasting Improvement Act, which as written, would have required the U.S. National Oceanic and Atmospheric Administration (NOAA) to shift its focus away from studying the climate crisis. The apparent hope was that less information would lead to fewer calls for action.[23] The bill never passed, but that didn't make its introduction any less unsettling.

In 2015, GOP lawmakers took aim at the Pentagon's and CIA's budgets because Republicans didn't want officials from either one conducting any research related to climate change as a national security issue. "The Department of Defense and the Central Intelligence Agency, two of the most important agencies in our national security apparatus, currently spend part of their budget studying climate change," the party's budget blueprint said.[24]

It wasn't an observation; it was a complaint. The text was included in a section of the budget plan headlined "Eliminating Waste."

Writing for the *New Republic,* Rebecca Leber commented, "According to House Republicans, study of the climate's impact on national security fits under 'examples of areas where there should be room to cut waste, eliminate redundancies, and end the abuse or misuse of taxpayer dollars.' But this is hardly a misuse of funds, considering that the Pentagon warned as far back as the George W. Bush administration that climate change is a threat to national security. The GOP simply wants the Pentagon to ignore the problem."[25]

It was emblematic of the Republican Party's antigovernance

posture. As official evidence of climate change and its effects reached Capitol Hill, GOP lawmakers were seemingly presented with a choice: act on the information or ignore it. But as the party became more comfortable with a post-policy approach, many Republicans found a third option: take deliberate steps to ensure policy makers have less information. It was part of a ridiculous what-we-don't-know-can't-hurt-us style of governance, which the GOP was a little too eager to embrace.

In March 2015, for example, Senator Ted Cruz of Texas, one of Congress' least-liked members, lobbied NASA to adopt a "more space, less Earth" strategy that would downplay the agency's climate research.[26] The same month, House Republicans unveiled a bill to "improve" the U.S. Environmental Protection Agency's Science Advisory Board by legally prohibiting it from considering research on climate science from a variety of sources, including the U.S. Global Change Research Program's National Climate Assessment, which is intended to serve as the federal government's definitive report on the climate crisis.[27]

Journalist Dave Roberts, writing on the website Grist, which covers environmental and energy issues, explained soon after that under the proposal, the EPA, when considering what to do about carbon pollution, would not be allowed to "consider what America's best scientists have concluded about it, what an international panel of scientists has concluded about it, how the federal government has officially recommended calculating its value, or the most comprehensive solutions for it."[28]

It was one thing for a political party to fail to respond responsibly to a crisis; it was something else, though, to see Republicans show such overt hostility toward reliable information about the crisis.

A governing party recognizes the importance of rigorous policy making, evaluates evidence before and after trying to implement ideas, starts with a question and works toward an answer rather than the other way around, and is swayed by reason and data. On climate policy, the GOP has spent too many years making clear that it is unwilling to even pretend to be a governing party.

It was around this time that Jim Inhofe, who chaired a committee overseeing environmental policy, did something never before seen on a congressional floor: he brought a good-sized snowball with him to the Senate and threw it to an aide.[29]

"You know what this is?" the Oklahoman asked on the Senate floor. "It's a snowball just from outside here. So it's very, very cold out. Very unseasonable."

It was February in Washington, DC. Snow and cold temperatures were not uncommon, and such conditions were hardly evidence against global warming. How anyone could expect federal lawmakers from both parties to work with someone like Inhofe on effective climate legislation was a total mystery.

The senator has long been a uniquely odious voice on these issues, but Inhofe's GOP colleagues were hardly better prepared for credible policy making. The month after the "Senator Snowball" incident, Ted Cruz made the case that his status as a climate denier made him a modern-day Galileo. Seriously.

As part of an interview with the online publication Texas Tribune, the Republican senator said of his critics, "What do they do? They scream, 'You're a denier!' They brand you a heretic. Today the global warming alarmists are the equivalent of the flat-earthers. It used to be [that] it is accepted scientific wisdom the Earth is flat, and this heretic named Galileo was branded a denier."[30] Cruz did not appear to be kidding.

Putting aside the fact that the Italian scientist was born after people knew already that the planet was round—it was his views on heliocentrism that made Galileo Galilei controversial in the sixteenth century—Cruz's argument was untethered to logic.[31] By his reasoning, anyone who rejects the consensus of the scientific canon on anything should be seen as a rebellious hero worthy of veneration.

Others in the GOP found value in trying to dodge the issue in a substance-free way. If rejecting the evidence made Republicans look like kooks and charlatans, and embracing the evidence offended their party's far-right base, many leading GOP officials decided the smart move would be to publicly question their own credentials instead of doing real work.

John Boehner, after spending years reading from the climate deniers' script, told reporters in May 2014, "Listen, I'm not qualified to debate the science over climate change."[32] A few days earlier, Republican governor Rick Scott of Florida was asked if he accepted the science surrounding anthropogenic global warming. "I'm not a scientist," he replied.[33]

A month before the 2014 midterm elections, Mitch McConnell was asked what it might take to persuade him to acknowledge the importance of the climate crisis. "I'm not a scientist," the Republicans' Senate leader said.[34]

It was odd to hear so many prominent voices in the party effectively declare that they had nothing of value to add to the fight over one of the world's most important policy challenges. Powerful GOP officials, in positions of elected authority, dealt with questions about an ongoing crisis by saying, in effect, that they didn't know what they were talking about. That was, to be sure, true, but it was also unhelpful.

What's more, the "I'm not a scientist" line seemed to direct those with questions to actual scientists—who in this case were the same people begging Republicans to deal with climate change like responsible adults.

The question was never whether lawmakers such as Boehner and McConnell had scientific expertise, but rather why they cared so little about real scientists' judgments.

Making matters worse, the defense, to the extent that it could be described as a defense, was hilariously narrow. Asked about, say, the ongoing war in Afghanistan, it didn't occur to GOP leaders to say, "I'm not a general." Asked about net neutrality, they never replied, "I'm not a computer scientist." During the jobs crisis, no Republican official ever passed on questions by throwing up his or her arms and declaring, "I'm not an economist."

Barack Obama grew exasperated by the party's talking points, telling an audience in June 2014: "I'm not a scientist, either, but I've got this guy, [White House Office of Science and Technology Policy Director] John Holdren, he's a scientist. I've got a bunch of scientists at NASA, and I've got a bunch of scientists at EPA. I'm not a doctor, either, but if a bunch of doctors tell me that tobacco can cause lung cancer, then I'll say, 'Okay!' It's not that hard. I'm not a scientist, but I read the science."[35]

Those elected to positions of great influence and power aren't expected to be experts in every discipline or area of public policy. The authorities are expected to have aides, researchers, and access to experts whose job it is to advise officials on every issue under the sun, including science.

Leaders such as McConnell and Boehner were required by the nature of their responsibilities to make judgment calls based on the best available evidence. When it came to climate change, the evidence told them to act. It also told them failure to act would intensify the crisis and make the consequences more severe. Republicans were presented with a challenge. They made a choice not to care.

Though it hardly seemed possible, Donald Trump managed to care even less.

Before formally launching his electoral career, Trump was unrestrained in his condemnations of climate science, dismissing the evidence as "a total con job," "pseudoscience," and a "hoax." Pointing to cold weather in January 2014, he implored the public to abandon any further discussion of the issue, calling climate change "bullshit."[36]

At least, that was Trump's posture when considering the crisis and its possible impact on Americans. His approach was quite a bit different, however, when it came to protecting his own properties: Politico reported in May 2016 that erosion and rising sea levels were such a threat to Trump International Golf Links, in Doonbeg, Ireland, on the North Atlantic coast, that the Republican's business launched an effort to build a sea wall, explicitly pointing to "global warming and its effects" to justify its necessity.[37]

As part of the same effort, the Trump Organization argued that Ireland's official data on soil erosion wasn't dire enough—because the effects of climate change were likely to be even more severe than Irish officials expected. The need for immediate preparatory steps, the Trump Organization insisted, was "unavoidable."

Donald Trump the climate-change-denying candidate paid no attention to Donald Trump the climate-change-accepting resort owner, even though the latter was correct.

During one of the 2016 presidential general-election debates, Hillary Clinton reminded the audience that Donald Trump "thinks that climate change is a hoax perpetrated by the Chinese. I think it's real." The Republican quickly contested the claim, insisting, "I did not. I did not. I do not say that."[38]

He was obviously lying, and after the event, the HuffPost's Sam Stein asked a variety of Trump campaign surrogates for their reactions, including Representative Marsha Blackburn. At first, the Tennessee Republican claimed that she didn't have her glasses and therefore couldn't read the Trump tweet proving his deception. After having the tweet read to her, Blackburn eventually said, "I do not believe in climate change. I think the Earth is in a cooling trend."[39]

Reminded that nine of the hottest years in the history of the planet had been recorded in the previous decade, the Tennessee lawmaker insisted once again, "We have also seen [in] the past ten years a little bit of a cooling."

In January 2019 Blackburn was sworn in as a U.S. senator. Republican leaders quickly put her on the committee overseeing science policy.

Trump's ignorance didn't interfere with his election, and on his fourth day as president, he signed an executive order in support of the Keystone XL project, a controversial oil pipeline, despite the White House's having done practically no substantive analysis on the issue. The absence of due diligence was so complete that before the presidential decision, no one in the West Wing had spoken to the State Department experts who'd spent years researching the project from every angle.[40]

Trump proceeded to boast about job creation surrounding the pipeline, but according to the State Department's analysis, the Keystone project would be responsible for all of thirty-five permanent, full-time American jobs. Despite all the hype, Bloomberg News's Josh Green reported that the jobs total would be "about what you'd get from opening a new Denny's franchise."[41]

That didn't faze the project's Republican champions. Neither did the evidence that it wouldn't make much of a difference for consumers at the gas pump, nor the evidence about the environmental hazards associated with the tar sands mining process that made this pipeline project more controversial than most.

By all appearances, by 2017, GOP officials no longer thought of the Keystone XL project as just a pipeline. It took on totemic value for the party: Republicans came to believe Keystone was their way to exert dominance over Democrats and everyone else who dared to value environmental considerations over the demands of the fossil-fuel industry.

At the same time, Trump dispatched Scott Pruitt, the former attorney general of Oklahoma, to take the reins at the Environmental Protection Agency—a federal department Pruitt had repeatedly fought during his tenure in state government. The rationale behind the personnel choice seemed obvious: the far-right Republican was tapped for the job in order to steer the EPA in a direction polluters would like.

Pruitt's work at the agency became a case study in why conservatives should be among those discouraged by the post-policy shift in Republican politics. After all, Trump's EPA chief, before being forced to resign following an avalanche of corruption allegations the following year,[42] was hopelessly incompetent and too unfamiliar with the functions of government to execute his plan effectively.

When Pruitt's office completed a court filing on vehicle emissions, for example, the *New York Times* noted it was "devoid of the kind of supporting legal, scientific, and technical data that courts have shown they expect to see when considering challenges to regulatory changes."[43]

Richard Lazarus, a professor of environmental law at Harvard University, explained, "In their rush to get things done, [Pruitt and his team are] failing to dot their i's and cross their t's. And they're starting to stumble over a lot of trip wires. They're producing a lot of short, poorly crafted rulemakings."

Ironically, during Pruitt's tenure at the EPA, environmentalists took some solace in the Republican's ineptitude. Had the Trump administration cared more about how government and policy making worked, the results would have been worse for the public.[44]

Too often, however, Republican indifference toward substance was more menacing. This was clearly the case when Trump turned his attention to withdrawing the United States from the Paris climate accord, which American officials had helped negotiate in 2015. The president was convinced the international agreement was a "bad deal," though he had great difficulty explaining what he didn't like about it.

As Trump's formal declaration grew closer, the White House faced intense pressure from U.S. allies and American business leaders who thought the president might be open to persuasion if only they could make a compelling enough argument. However, their lobbying was predicated on the assumption that Trump was a mature policy maker who cared about responsible governance and would base a decision on the magnitude of evidence and reason.

Efforts to move Trump failed because those assumptions were

wrong. Kellyanne Conway, a prominent presidential adviser and spokesperson, conceded that when it came to the climate accord, "He started with a conclusion."[45] She did not intend that as criticism, which only helped reinforce concerns about how policy decisions came together in the Trump White House.

When the president appeared in the Rose Garden in June 2017 and thanked attendees for being on hand for the event in which he announced his intention to withdraw from the international climate agreement, the first sentence out of his mouth was, "I would like to begin by addressing the terrorist attack in Manila." It was an inauspicious beginning: there had been no terrorist attack in Manila. As Trump probably should have known, there'd been a casino robbery in the Filipino capital earlier in the day, but it hadn't been pulled off by terrorists.[46] (Officials in the White House Situation Room began laughing after they heard the president's comment.[47])

The speech went downhill from there. Despite reading from scripted remarks, Trump left no doubt he knew very little about what the Paris climate accord was, how it was structured, or even how it was supposed to work in practice. The president knew he didn't like the agreement, but in true post-policy fashion, he didn't know why.

At times, it seemed as if Trump were withdrawing from a pact that existed only in his imagination. In his remarks, for example, the Republican condemned the "onerous energy restrictions" imposed on the United States by the Paris accord. No such restrictions existed.

He added, "As of today, the United States will cease all implementation of the nonbinding Paris accord and the draconian financial and economic burdens the agreement imposes on our country." The presidential contradiction went unexplained: accords can be "nonbinding," or they can impose "draconian" burdens, but they can't do both.

Trump went on to say that under the Paris agreement, "China

will be allowed to build hundreds of additional coal plants. So, we can't build the plants, but they can, according to this agreement. India will be allowed to double its coal production by 2020." This didn't make sense: nothing in the climate accord specified which countries could or could not build coal plants—or any other sources of energy. The policy was about countries reaching reduced-emission targets, not defining in detail sources of domestic power.

On and on it went, tiring fact-checkers who struggled to keep up with the avalanche of misstatements. Writing for Vox, Dave Roberts said of the remarks, "It is a remarkable address, in its own way, in that virtually every passage contains something false or misleading. The sheer density of bullshit is almost admirable, from a performance art perspective."[48]

Looking ahead, Trump added, "We're getting out, but we will start to negotiate, and we will see if we can make a deal that's fair." There were no subsequent talks toward a new agreement, fair or otherwise, in large part because there were no other countries with whom the president was prepared to talk.

Indeed, at the heart of Trump's address was a paranoid fantasy involving nefarious foreign rascals trying to pull one over on American dupes. "The Paris Agreement handicaps the United States economy in order to win praise from the very foreign capitals and global activists that have long sought to gain wealth at our country's expense," he argued, adding, "The fact that the Paris deal hamstrings the United States, while empowering some of the world's top polluting countries, should dispel any doubt as to the real reason why foreign lobbyists wish to keep our magnificent country tied up and bound down by this agreement: it's to give their country an economic edge over the United States."

None of this was in any way true, but it did offer a peek into Trump's genuine rationale, which plainly had very little to do with public policy. Michael Grunwald's explanation in Politico rang true: "It was about extending a middle finger to the world, while reminding his base that he shares its resentments of fancy-pants elites and smarty-pants scientists and tree-hugging squishes who look down on real Americans who drill for oil and dig for coal."[49]

Those who tried to persuade Trump to be more responsible never stood a chance. Told that his decision would leave his own country isolated and reviled, the Republican saw a selling point. Told that his rationale for abandoning an international agreement bordered on gibberish, the president embraced rhetoric that sounded more like refrigerator poetry than a policy analysis.

The move wasn't about substance; it was about making Trump feel better about an issue he didn't even want to understand.

In the run-up to the formal White House declaration, Vice President Mike Pence appeared on Fox News and seemed confused about the nature of the climate fight itself. "For some reason or another, this issue of climate change has emerged as a paramount issue for the Left in this country and around the world," Pence said in June 2017.[50]

In 1990 the UN's Intergovernmental Panel on Climate Change published one of the first major reports warning the public the world over about the burgeoning crisis. More than a quarter century later, Mike Pence realized that people around the world considered the climate crisis important, but he wasn't sure why they were making such a fuss.

The administration's willful ignorance did not improve. The month after the White House speech on the Paris accord, Trump appeared at a campaign rally in Phoenix, where he spoke with great pride about part of his energy agenda. "We've ended the war on beautiful, clean coal," the president said. Apparently hoping to help his audience understand the policy, he added, "They're going to take out clean coal—meaning, they're taking out coal, they're going to clean it."

The idea behind clean coal—a phrase touted by the industry since the Bush-Cheney era—is a specific technique in which coal-powered plants try to capture emissions and bury the pollution underground via pipelines. The process is often referred to as "carbon capture and sequestration." It does not involve anyone "taking out coal" and then "cleaning it." Even when talking about policies he ostensibly likes, Trump often doesn't know what he's saying—and doesn't care to find out.

The Republican embraced the idea of clean coal as a candidate, endorsing the idea during one of the general-election debates, and then incorporating the approach into his administration's energy policy. At no point, however, did it occur to Trump to learn what clean coal is. Even normal presidents can't be expected to have vast subject-matter expertise on every issue, but this president broke new ground, lacking basic familiarity with his own ideas, despite years of rhetoric on the subject.

In Trump's mind, he wanted to support coal for political reasons: he thought *clean* sounded like a complimentary adjective, so he started referring to *all* coal as clean coal because it seemed like a nice thing to say. In 2018, indifferent to the industry term of art, he told reporters, "I just left Montana, and I looked at those trains, and they're loaded up with clean coal—beautiful clean coal."

In some cases, Trump even tried to convince people that coal is unbreakable. At a campaign rally in 2018, ahead of the midterm elections, the Republican told supporters, "We love clean, beautiful West Virginia coal. We love it. And you know that's indestructible stuff. In times of war, in times of conflict, you can blow up those windmills. They fall down real quick. You can blow up those pipelines. . . . You can do a lot of things to those solar panels. But you know what you can't hurt? Coal."

As exasperated fact-checkers at the *New York Times* noted, "If coal itself were truly indestructible, you couldn't mine or burn it."[51]

Soon after, Trump sat down with the Associated Press, which asked about the climate crisis and his passivity toward the issue. The president replied, "My uncle was a great professor at MIT for many years. Dr. John Trump. And I didn't talk to him about this particular subject, but I have a natural instinct for science, and I will say that you have scientists on both sides of the picture."[52]

Among the problems was his apparent confusion over the meaning of the word "instinct." Some people know a great deal about science, others don't, but the latter group cannot instinctively know scientific details without a base of information and experience upon which to draw.

Asked about the value of a high-rise in Midtown Manhattan, Trump would be well positioned to give a credible instinctive answer. Asked about the effects of anthropogenic climate change, the instincts of a former reality-show personality are unpersuasive—his uncle's career in academia notwithstanding.

The president, however, has long assumed that baseless assumptions resulting from superficial news consumption and assorted motivated guesses constitute "a natural instinct for science." Those instincts were on full display in early 2019, when he started talking in more detail about clean-energy alternatives, with a special emphasis on his opposition to wind power. Trump told a Michigan audience about the energy horrors they would have experienced in the event of a Hillary Clinton presidency.

"You would be doing wind, windmills, and if it doesn't blow, you can forget about television for that night," the president said to laughter. Imagining a family conversation, he added, "'Darling, I want to watch television.' 'I'm sorry, the wind isn't blowing.'" He concluded that the audience should trust that he knows what he's talking about. "I know a lot about wind," the Republican said.

There were gaps in his knowledge. The Department of Energy's Office of Energy Efficiency and Renewable Energy had a "Frequently Asked Questions About Wind Energy" page on its website, responding to Americans concerned about electrical supplies during calm skies. The agency pointed to ample research that proved the domestic power grid could handle "variable renewable power without sacrificing reliability."[53] It's not one of the more complicated truths to understand about energy policy. Similarly, families who live in homes with solar panels do not sit in darkness after sundown.

Unembarrassed, Trump took a similar message to the National Republican Congressional Committee in April 2019, warning officials about the carcinogenic qualities of wind turbines. "They say the noise causes cancer," the president argued, failing to identify who "they" were. [54] He also didn't try to reconcile his assertion with his "natural instinct for science."

As those who took the substance of governing seriously likely knew, there is no evidence whatsoever linking wind power and cancer.[55] There is, however, abundant evidence connecting cancer to coal power, which the president was eager to champion.

A day later, reporters pressed Mercedes Schlapp, the White House's director of strategic communications, for some kind of explanation as to why the president said during nationally televised comments that wind power can cause cancer. "I don't have an answer to that," Schlapp replied. "I don't, I—I—I don't have an answer to that. Yeah, I really don't have information on that."[56]

By most accounts, Trump's hostility toward wind power stemmed from his struggling Scottish golf resort and his belief that eleven nearby wind turbines detracted from his business's views.[57] The Republican brought these attitudes with him to the Oval Office, reinforcing suspicions that the American president, even while in office, was focused principally on personal profit, not responsible policy making for the good of the country.

At times that meant ignoring scientific evidence, and at times it meant ignoring officials in the Trump administration. In January 2019 Director of National Intelligence Dan Coats—appointed by Trump and confirmed by Senate Republicans—released his office's annual *Worldwide Threat Assessment* report. Its findings, putting climate change in a national security context, were startling.

After noting a series of horrifying climate hazards, Coats wrote, "Extreme weather events, many worsened by accelerating sea level rise, will particularly affect urban coastal areas in South Asia,

Southeast Asia, and the Western Hemisphere. Damage to communication, energy, and transportation infrastructure could affect low-lying military bases, inflict economic costs, and cause human displacement and loss of life."

This came just two weeks after Trump's Pentagon issued a related report, noting that climate change is a global security threat. "The effects of a changing climate are a national security issue with potential impacts to Department of Defense missions, operational plans, and installations," it warned.[58]

None of this, however, stopped the president from marveling at cold temperatures throughout the Midwest in a tweet from January 2019. "What the hell is going on with Global Waming [*sic*]?" Trump wrote. "Please come back fast, we need you!"[59]

Officials in the Office of the Director of National Intelligence and the Pentagon had just told the president what was going on with climate change, but Trump had no interest, even with regard to national security implications.

In Congress, the debate over constructive steps to address the climate crisis wasn't much better. Former secretary of state John Kerry, who helped negotiate the Paris climate accord, testified during a Democratic-led House Committee on Oversight and Reform hearing on the need for action on the issue.

When it was his turn to ask questions of the witnesses, Republican congressman Thomas Massie of Kentucky thought it'd be a good idea to inquire about Kerry's undergraduate "science degree" from Yale University. The Democrat replied by reminding the congressman that his bachelor's degree was in political science.

"How do you get a bachelor of arts in a science?" Massie asked.[60] Kerry, who'd represented Massachusetts in the Senate for twenty-eight years before replacing Hillary Clinton as secretary of state, started to explain the nature of a liberal arts education when the Kentucky lawmaker interjected, "So, it's not really science. I think it's somewhat appropriate that someone with a pseudoscience degree is here pushing pseudoscience in front of our committee today."

Rolling Stone cited Massie's line of questioning as possibly "the dumbest moment in congressional history," and while that was probably hyperbole, the magazine wasn't alone in its amazement.[61] At the congressional hearing itself, Kerry, clearly exasperated, asked, "Is this really serious? Is this really happening here?"

He added later on Twitter, "It's almost as if someone said, 'Congress has hit rock bottom,' and Massie replied with, 'Hold my beer.'"[62]

Those who watched the exchange and assumed the GOP congressman was dumb were mistaken. Massie, in fact, had earned two engineering degrees from MIT. The problem was that he appeared to be acting in bad faith, rejecting the science and refusing to do any work toward a solution.

The exchange between the two brought the scope of the post-policy problem into sharper focus: How can the legislative branch of the world's preeminent superpower respond to crises such as climate change when educated and experienced Republicans like Thomas Massie are more comfortable engaging in embarrassing antics than assessing the evidence? How does policy making happen at all when officials from one party decide they don't want to believe the facts presented to them?

Though it may have seemed impossible, the political debate, such as it was, took a more farcical turn following the 2018 midterm elections, after which some congressional Democrats advocated in support of an agenda called the Green New Deal. It was a deliberately vague platform intended to direct attention toward aggressive climate benchmarks.

To put it mildly, the blueprint was not well received by Republicans. Representative Rob Bishop of Utah told reporters at a

March 2019 press conference, "For many people who live in the West, but also in rural and urban areas, the ideas behind the Green New Deal are tantamount to genocide. That may be an overstatement, but not by a whole lot."[63] (Two months earlier, GOP leaders named Bishop as the top Republican on the House Natural Resources Committee.)

The Utahan wasn't alone. At the 2019 Conservative Political Action Conference (CPAC), one of the year's biggest annual confabs for Republicans and their ideological allies, Ted Cruz told attendees that the Green New Deal was proof that Democrats intend to "kill all the cows." Sebastian Gorka, briefly a controversial aide in the Trump White House, added, "They want to take away your hamburgers. This is what Stalin dreamt about but never achieved."[64]

Gorka's curious understanding of Soviet history aside, the Green New Deal made no reference to cows or burgers, although the office of a freshman Democratic representative from New York, Alexandria Ocasio-Cortez, mistakenly and prematurely distributed a fact sheet about the blueprint that jokingly said the resolution "set a goal to get to net-zero, rather than zero emissions, in ten years because we aren't sure that we'll be able to fully get rid of farting cows and airplanes that fast."

This was obviously intended to be funny, and the congresswoman's office quickly retracted the materials. GOP officials nevertheless seized on the joke as evidence of a widespread anticow Democratic agenda.

Ocasio-Cortez told the *New Yorker*'s David Remnick that "it's hard for the Republicans to refute the actual policy on its substance. They resort to mythologizing it on a ludicrous level."[65]

There was ample room for a debate about the role of the agricultural industry, including livestock, in addressing global warming, and what role decreased meat consumption might play as part of a multifaceted solution. But GOP officials much preferred a post-policy fight against a cartoonish vision of ambitious climate targets.

For his part, Mitch McConnell scheduled a show vote on the Green New Deal platform, despite his fierce opposition to it, because he hoped to play a little political game: Senate Democrats could either vote for the ambitious agenda and invite attack ads, or vote against it and disappoint parts of the party's base.

But it was the debate on the measure that ended up standing out. Republican senator Mike Lee of Utah suggested he'd identified a new way to address global warming that hadn't been considered before. "Climate change is an engineering problem—not social engineering, but the real kind," he claimed. "It's a challenge of creativity, ingenuity, and technological invention. And problems of human imagination are not solved by more laws, but by more humans! More people mean bigger markets for innovation. More babies mean more forward-looking adults—the sort we need to tackle long-term, large-scale problems."[66]

In other words, Lee endorsed a probability solution to the climate crisis:

1. Some people need to solve the problem.
2. If there are more people, it increases the odds that some of them might find a solution.

Part of the problem with this deeply strange and fundamentally unserious pitch was that population increases make it more difficult to address the climate crisis, not less. Just as importantly, the Utahan made it sound as if the solution to the climate problem was elusive, but that's never been true: responsible officials have known for quite a while what's necessary to address the crisis. The need for more people in the overall population pales in comparison with the need for more people in positions of authority who are committed to implementing effective policies.

The climate crisis is not a black box with mysterious contents. As difficult as the solutions are, the questions of how to address the climate emergency already have answers. The senator and his cohorts may prefer to ignore the solutions, but to pretend they don't exist—or that we need a vastly larger population to help discover them—is ridiculous.

His bizarre speech notwithstanding, Lee has been commended as one of the more intellectual members of the Senate Republican Conference. It led *New York*'s Jonathan Chait to note after the senator's climate remarks, "What makes this praise so damning is that it's all probably true. This is a party whose members have declared that the existence of snow in February refutes climate science, and that rising sea levels are caused by a buildup of dirt on the ocean floor. Lee's speech at least doesn't challenge the fact that climate change is a real phenomenon. He probably is one of the smartest Republican legislators in Washington! Which is like being the most physically intimidating baby in the stroller park."[67]

McConnell's resolution failed with zero votes. The Democratic minority, loath to play along with the pointless partisan game, voted "present."[68]

CHAPTER 5

"A Series of Hasty Unplanned, Unexamined Decisions"

Foreign Policy

AFTER DECADES IN WHICH Republicans saw foreign policy as one of their signature issues, the party was at a crossroads after Barack Obama's presidential inauguration. The neoconservative hawks that dominated GOP thinking in the Bush-Cheney administration had been rejected by the American mainstream, rebuffed by allies, and discredited by the real-world results of their failed agenda.

Indeed, as 2008 came to a close, the United States found a landscape littered with avoidable international tragedies. The country was mired in two Middle Eastern wars, alienated from traditional allies, tarnished by a military torture scandal, and suffering from diminished respect and credibility on the global stage. The outgoing Republican administration was bequeathing a series of international disasters that were difficult to measure and even harder to address.

The Western world, having grown weary of Bush and his failures, was comfortable turning instead to a young Democratic senator to claim the mantle of international leadership and begin the work of cleaning up his predecessors' messes, which was no small endeavor. Repairing relationships, restoring the nation's stature, and taking steps to bring some semblance of stability to conditions in Iraq and Afghanistan posed a daunting challenge.

The Republican Party had a respected group of elder statesmen and stateswomen with extensive experience and credibility on international affairs, but they were gradually exiting the stage. As their GOP moved sharply to the right, the party's old-guard establishment—generally comfortable with bipartisanship on matters of global import—saw its influence waning, even among congressional Republicans who might be expected to seek its guidance.

Even the fundamental issue that recognizes the inherent value of a serious foreign policy seemed unresolved. The United States has a unique leadership role on the international stage, and a responsible vision is a necessity both for global stability and rules-based order.

When the world's preeminent superpower can no longer be counted on to approach international affairs in a mature fashion, there are real consequences—diplomatically, politically, economically, and militarily. Rules break down, institutions lose their influence, rogue actors begin to see opportunities, and pillars of global strength start to wobble. For much of the modern era, there was a broad, bipartisan consensus on this vision, but for contemporary Republicans, foreign policy assumptions were suddenly unreliable.

In early 2009 the GOP's uncertain vision created a fresh challenge for the incoming Democratic administration. Obama intended to pursue international priorities with fairly broad support and international legitimacy, while trying to appeal to a Republican Party that no longer had a discernable foreign policy of its own.

In the months and years that followed, GOP officials largely removed the ambiguities and made the process quite simple: the Republican Party's foreign policy would be shaped entirely by its opposition to whatever Barack Obama was for.

In his first week as president, Obama initiated a process to close the military detention facility at the Guantanamo Bay Naval Station in Cuba. It was a provision from the Democrat's successful campaign platform; it was a goal that had already been endorsed by his Republican rival in the 2008 campaign; and the idea enjoyed the support of U.S. military leaders. White House officials basically saw closing the prison—by most fair measures, a stain on our national character—as low-hanging fruit on the new administration's to-do list. This simple change wouldn't erase the recent history of torture and indefinite detention policies, but it would be a step in a defensible direction.

Finding substantive reasons to keep the Guantanamo Bay prison open was difficult. There were already a variety of maximum-security facilities on American soil housing convicted terrorists, as well as a judicial system adept at handling their cases. The national-security risks associated with closing the Cuban facility were nonexistent, and, according to counterterrorism experts, the national-security benefits were real. This was, for all intents and purposes, a no-brainer.

The Republican Party, however, didn't see it that way, and its response to the president's move was nothing short of hysterical. House GOP leaders quickly threw together television advertising suggesting that Obama was inviting another 9/11, and House Minority Leader John Boehner introduced the Keep Terrorists Out of America Act, intended to block the White House's policy. It picked up 169 cosponsors in short order.

Republican Kit Bond of Missouri, the ranking member of the Senate Intelligence Committee, delivered a national address in May 2009 declaring that Americans "have a right to know if President Obama plans to send any of these terrorists to their communities."[1] Soon after, Senator James Inhofe felt comfortable telling his Oklahoma constituents that Obama was "obsessed with turning terrorists loose in America."[2]

There was ample room for a substantive discussion over federal detention policies and the legal fate of suspected terrorists, but for

Republican lawmakers, a careful examination of the facts would have served no useful purpose. The party saw an opportunity to scare the daylights out of the public and undermine public support for the new Democratic president, who was still riding high with a strong approval rating.

The idea of governing in response to a hearty policy dialogue was never considered. In fact, Democrats and Republicans spent months talking past one another: the former would point to extensive evidence about American facilities already doing fine work imprisoning terrorists, and the latter would say evidence and mindless demagoguery don't mix. The GOP's post-policy tactics were in full effect.

Obama and his team hoped that the support of high-profile military leaders, who had no use for mindless scare tactics, might help turn the tide. Defense Secretary Robert Gates, appointed to the position by George W. Bush, was fully on board with closing the Guantanamo prison, as was Admiral Mike Mullen, the chairman of the Joint Chiefs of Staff, who was also appointed by Bush. They were soon joined by former secretary of state Colin Powell, another Bush administration veteran, and retired general David Petraeus, whom Bush had tapped to lead U.S. Central Command, overseeing military operations in the Middle East.

There may have been a point at which Republicans would have valued the national-security judgments reached by U.S. military leaders—especially a contingent that featured Powell and Petraeus—but so long as Obama was in the Oval Office, and the brass agreed with his agenda, the GOP considered their judgments irrelevant.

As ridiculous as the fearmongering was, it also worked like a charm: public anxieties led lawmakers, including many intimidated Democrats, to reject efforts to close the prison at Guantanamo Bay. U.S. military leaders continued to speak out—even retired Marine Corps major general Michael Lehnert, the first commander of the Guantanamo facility, endorsed its closure—but Republicans left no doubt that they would not be swayed by reason.[3] This wasn't a policy debate; it was a political campaign, and the GOP was winning.

Over the course of his presidency, Obama made incremental progress toward his goal, largely through diplomatic efforts that focused on finding other countries willing to take the detainees. The facilities' population peaked in 2003 with 680 prisoners, and when George W. Bush left office, it was down to 242 prisoners. The day Obama exited the White House, that total had shrunk to just 41.

Some in the Democratic administration hoped fiscal sanity would eventually be part of the conversation: the smaller the number of detainees, the harder it would be to justify the massive expense of keeping open a detention facility that housed so few people. Even if congressional Republicans were inclined to ignore every other consideration, the hope was that GOP lawmakers would at least care about wasteful taxpayer spending. Despite a price tag of $13 million per prisoner, the party wouldn't budge.[4]

The GOP's approach to terrorist threats was routinely detached from substance. Indeed, throughout Obama's two terms, Republicans were preoccupied to the point of obsession with the White House's rhetorical choices related to dangerous militants, far more than the administration's policies themselves.

In January 2010, in the wake of a failed terrorist attempt, Republican congressman Peter King of New York appeared on ABC's *Good Morning America* and was asked to name a specific recommendation he'd make to the president to improve national security. King, the ranking Republican on the House Security Committee, replied quickly, "I think one main thing would be to—just himself to use the word *terrorism* more often."[5]

For the better part of a decade, this was a staple of GOP thinking on the subject, with one Republican after another expressing outrage over the fact that the Democrat had no use for the phrase "war on terror," as if its mere utterance would magically keep Americans safe. In the immediate aftermath of terrorist violence in Paris in 2015, Senate Majority Whip John Cornyn of Texas argued that proper conservative word choice was "the first step to actually dealing with [the security threat] on a realistic basis."[6] Around the same time, Senator Richard Burr, newly chosen to lead the Senate Intelligence Committee, said his "only concern" was Obama's rhetoric.[7]

A year later, Donald Trump took the line to its illogical extreme, publishing a tweet that read, "Is President Obama going to finally mention the words radical Islamic terrorism? If he doesn't, he should immediately resign in disgrace!"[8]

The party chose to overlook the fact that it was the Bush-Cheney White House that had issued rhetorical guidelines, titled "Words That Work and Words That Don't: A Guide for Counterterrorism Communication." The directive urged officials to stop describing militants as "jihadists" and to drop phrases such as "Islamo-fascism" altogether. "It's not what you say but what they hear," the memo said in bold lettering.

Karen Hughes, a top adviser to George W. Bush, explained later, "We ought to avoid the language of religion. Whenever they hear 'Islamic extremism, Islamic jihad, Islamic fundamentalism,' they perceive it as a sort of an attack on their faith. That's the world view Osama bin Laden wants them to have."

Throughout Obama's presidency, Republicans had no use for the Bush-era guidance, dismissing it as weak and unrealistic, evidence be damned.

When GOP officials weren't arguing that word choice was central to an effective counterterrorism strategy, many in the party were inclined to leave the substance of the issue to others. For example, as the Obama administration intensified its offensive against the Islamic State of Iraq and the Levant—the terrorist network generally known as ISIS—former Republican representative

Jack Kingston explained why many congressional Republicans had little interest in formally authorizing the mission.

"A lot of people would like to stay on the sideline and say, 'Just bomb the place and tell us about it later,'" the Georgian said.[9] He added, in reference to Republicans' attitudes, "We like the path we're on now. We can denounce it if it goes bad, and praise it if it goes well, and ask what took him so long."

Kingston's candor was welcome, but it was still remarkable to hear a member of Congress describe such passivity on national security out loud and on the record.

The GOP's policy indifference dominated the era. In 2010 Obama and Russian president Dmitri Medvedev came to terms on the New Strategic Arms Reduction Treaty (New START), building on previous nuclear agreements between the two countries. The legacy of those policies was a source of international pride, and for good reason: the U.S.-Russian arms deals helped keep the world safe from the most dangerous weapons ever created. Officials in both countries had a vested interest in maintaining and building on that legacy.

The 2010 deal created new caps on the number of nuclear warheads in the countries' respective arsenals, limited the number of nuclear missile launchers, and expanded the existing treaties' inspections process. They were, by any fair measure, straightforward provisions that satisfied both nations' security needs.

Once the agreement was negotiated, time was of the essence: New START was the successor to the START I treaty, which had expired in late 2009. With the new deal completed in April 2010, it was widely assumed that U.S. and Russian officials would act quickly to approve and implement the treaty.

On Capitol Hill, where treaties require two-thirds support for ratification, the race was on to find sixty-seven U.S. senators who were willing to vote for the new agreement. Foreign policy experts assumed this would require very little effort. They failed to realize just how far the Republican Party had strayed from its status as a responsible governing party.

Almost immediately after the Obama administration announced the terms of the treaty, a series of GOP senators expressed deep skepticism for reasons they struggled to explain. Senate Minority Whip Jon Kyl led the opposition, though he hadn't bothered to do his homework: the Arizonan said in August 2010 that he assumed nuclear inspections would continue while deliberations were under way. "I thought we were just going to continue doing business as usual," the GOP leader said.

Had he done his due diligence before starting his obstruction campaign, the senator would have known better. Because START I had expired several months earlier, the United States, for the first time in nearly two decades, could no longer inspect Russian long-range nuclear facilities.[10] Reflecting on Kyl's policy negligence, the *Washington Post*'s David Broder responded soon after, "What a price to pay for ignorance."[11]

The more Republicans resisted, the more it seemed to put ratification of New START in doubt. Officials in Russia were "mystified," unsure what it was about the deal Republicans didn't like, but the bafflement wasn't limited to Moscow.[12] French ambassador Pierre Vimont and other diplomats sent cables home warning of the difficulties getting the U.S. Congress to approve the treaty. Vimont said officials in Europe were incredulous: "People ask us, 'Have you been drinking?'"[13]

The confusion was understandable, since international observers assumed that American officials would be principally concerned with policies that would benefit the United States. That would have been true if Republican lawmakers were members of a governing party instead of a party that had abandoned the pretense that the substance of public policy mattered. The Democratic president and his team negotiated the terms of the deal, and for most GOP senators, there was little else to discuss.

Republican opposition was not, however, entirely universal. Senator Richard Lugar of Indiana, the ranking member on the Senate Foreign Relations Committee, examined the terms of the treaty carefully and positioned himself as one of its most ardent champions. He was joined by prominent veterans of GOP foreign policy debates, including four former secretaries of state (Colin Powell, Condoleezza Rice, Henry Kissinger, and George Schultz) and three ex-senators (Chuck Hagel of Nebraska, Tennessee's Howard Baker, and John Danforth of Missouri). Chairman Mullen of the Joint Chiefs of Staff added that the treaty enjoyed "the full support" of the uniformed military. Even former president George H. W. Bush publicly endorsed the treaty.

In all, six former secretaries of state, five former secretaries of defense, seven former U.S. Strategic Command chiefs; former White House national security advisers from both parties, and nearly all former commanders of U.S. nuclear forces publicly urged the Senate to ratify New START. American allies quickly joined the chorus, calling on Republicans to be responsible and support the treaty.[14]

In response, most GOP senators simply shrugged. The calls weren't persuasive because the party wasn't listening.

Some of the details may have seemed familiar to observers with long memories. For example, during the Cold War, the right opposed every nonproliferation treaty with the U.S.S.R.[15] In 1972 the right also criticized Richard Nixon for opening relations with China, as well as for signing the U.S.-U.S.S.R. SALT (Strategic Arms Limitation Talks) treaty. Likewise, conservatives condemned the original START treaty, which was implemented in 1994, following the collapse of the Soviet Union. Though Ronald Reagan later reached near-deity status in conservative circles, even the iconic Republican president was branded an appeaser by far-right ideologues for trying to negotiate with the Soviets.

But in each of those instances, most mainstream Republicans leaders on Capitol Hill tried to keep the crackpots at arm's length and supported responsible diplomacy, even with American enemies. However, as the GOP descended deeper into a post-policy posture, the lines between the fringe and the mainstream blurred. The crackpots were helping steer the party.

The Obama White House approached the New START fight as a policy dispute and went to great lengths to try to address the party's concerns in good faith. Administration officials even traveled to Arizona in order provide Kyl, the number two Senate Republican and chief New START foe, with a series of detailed briefings on the treaty's merits.[16] The assumption was that substance and bulletproof arguments would eventually prove persuasive.

By November 2010, after twenty-nine meetings with Kyl and his team, the Arizonan said he wanted the Obama administration to agree to billions of dollars in new investments in nuclear modernization. Fine, the White House said, committing to spend $80 billion over the next decade.[17]

Officials in the West Wing thought they had a deal. Kyl opposed the treaty anyway.

As the debate on Capitol Hill came to a head, Dick Lugar, two years before Republican voters in his native Indiana would turn on him in a GOP primary, implored his colleagues, "Please do your duty for your country." Most senators did: New START was ratified with seventy-one votes. But by a two-to-one margin, GOP senators opposed the agreement, ignoring the pleas of a former Republican president, Republican Pentagon chiefs, Republican secretaries of state, and U.S. military leaders who'd served in Republican administrations.[18]

Writing for *Time,* Michael Crowley observed during the debate, "What you're seeing . . . is a struggle between the GOP's old-guard national security center and a resurgent, post-Iraq hawkish right wing."[19] There was certainly some truth to that. But the difference wasn't limited to style or ideology, and there was no real debate, per se, between the Republican Party's elder statesmen and their right-wing counterparts. The dividing line was one of perspective and priorities: the GOP's old guard cared about the details of the treaty and the security consequences of failure, while the party's far-right senators cared about saying no to Barack Obama. In the Venn diagram, there was little meaningful overlap between the circles.

New START was one of several treaties to struggle in the face of mindless Republican obstinance. In 2012, for example, the Senate took up the United Nations Convention on the Rights of Persons with Disabilities, a treaty that had already been ratified by more than 120 other countries, and the terms of which had been negotiated by the Bush-Cheney administration.

From an American perspective, the point of the treaty was to extend the benefits of our 1990 Americans with Disabilities Act to people around the world. When the Senate Foreign Relations Committee approved the treaty in July 2012, Chairman John Kerry explained that the proposal simply "raises the [international] standard to our level without requiring us to go further." Its ratification, at least initially, seemed obvious.

The right balked anyway. South Carolina senator Jim DeMint, for example, inexplicably said ratification of the treaty might expand abortion rights.[20] Former Republican senator Rick Santorum of Pennsylvania, by this time a political pundit on CNN, was an especially fierce critic of the measure, insisting that the treaty—which, again, would have required the United States to do literally nothing except wait for the world to follow the example we'd set already—represented "a direct assault" on conservatives.[21]

Retired former Senate majority leader Bob Dole, badly injured during his service in World War II, took on the UN Convention on the Rights of Persons with Disabilities as a personal cause, confident that he could apply his leadership skills to one last important measure. The Kansas Republican worked the phones, as members of the chamber he once led considered whether to ratify the treaty. The day of the vote, Dole made a rare in-person appearance.

As the *Boston Globe* reported, Dole, eighty-nine years old and just out of the hospital, "rolled in his wheelchair onto the Senate floor, all but daring senators to vote against him and, by proxy, anyone with a disability."[22]

It was a powerful display that members of his party were content to ignore. Senate Democrats supported the human rights treaty, but they were joined by only eight GOP lawmakers, leaving the measure far short of what was needed for ratification. A total of thirty-eight senators linked arms to defeat the pact—all of them Republican.

Opponents didn't have much of an argument against the measure, and, for the most part, they didn't even try to defend their position. Dan Drezner, a center-right professor of international politics at Tufts University, scrutinized the GOP arguments. Noting the gap between the party's claims and reality, he described the Republicans' case as "dumber than a bag of hammers."[23]

But Republicans' embrace of post-policy politics ignores assessments like these because the party no longer feels like it needs sound arguments to prevail. GOP senators knew that Democrats supported the treaty, and the far-right base did not. The fact that the policy enjoyed the backing of the party's old-guard statesman was again deemed irrelevant, which is why the debate was over long before the vote was held on the Senate floor.

A week earlier, the Senate also tried to ratify the Law of the Sea Treaty: an international legal framework, decades in the making, intended to establish international rules pertaining to marine trade, security, and natural resources. As the debate took shape, the Obama administration told senators the treaty would allow the United States to "secure mineral rights in a larger geographical area, would ensure freedom of navigation for U.S. ships, and would give the country better leverage for claims in the Arctic."

When the Senate Foreign Relations Committee kicked off the ratification process, it did so with a twenty-four-star hearing: a witness panel featuring six U.S. military officers with four stars apiece, each of whom told senators it was in Americans' interests to ratify the treaty.[24]

There wasn't anything especially partisan or ideological about the proposal, and just as with the New START debate, the treaty was endorsed by, among others, Colin Powell, Condoleezza Rice, Henry Kissinger, George Shultz, and yet another former secretary of state, James Baker, who served in both the Reagan and George H. W. Bush administrations, in addition to domestic business leaders, the State Department, the Pentagon, the Joint Chiefs, and U.S. Navy leaders.

In theory, it looked like another measure that could enjoy bipartisan backing. In practice, Senate Republicans derailed this treaty, too, pointing to paranoid fears from the GOP's right-wing base—amplified to great effect by Fox News—that the Law of the Sea proposal would lead to the United Nations superseding American laws.

Whether Republican lawmakers actually believed these bogus arguments was the subject of some debate, but GOP officials knew its party's activists were told the conspiracy theories were true, so the party felt as if it had no choice but to go along, substance be damned.

A year later, the United Nations General Assembly endorsed a proposal known as the Arms Trade Treaty by a vote of 154 to 3. While there were some abstentions, the only nations opposed to the ATT were Iran, North Korea, and Syria. The United States, under the Obama administration's leadership, voted with the majority, and several months later, Secretary of State John Kerry officially endorsed the treaty, beginning the ratification process.

Any time that gun policy is at the fore in the United States, it's safe to assume a contentious political process will soon follow, though in this case, the Arms Trade Treaty would have had no effect on the Second Amendment or domestic gun laws, and imposed no new restrictions on American consumers' access to firearms.

Rather, the point of the treaty was to set controls on the international gun trade. As Kerry explained after helping negotiate it, the ATT would "help reduce the risk that international transfers of conventional arms will be used to carry out the world's worst crimes, including terrorism, genocide, crimes against humanity, and war crimes."

Uninterested in those details or the text of the policy, Republican senator Rand Paul of Kentucky insisted that abiding by the treaty would lead to an international registry of American gun owners and "full-scale gun confiscation."[25] Senator James Inhofe added that the treaty was part of an effort to "sign away our laws to the global community and unelected UN bureaucrats."[26]

Arguments like these were demonstrably ridiculous, but they ultimately prevailed: Senate Republicans never allowed the ATT to even come up for a vote. (In 2019 the Trump administration unsigned the unratified treaty on behalf of the United States.)

The post-policy party helped usher in a post-treaty phase of American leadership. The GOP has long been aware of this, but it is also ambivalent about the dynamic's effects.

The clarity of the Republicans' antigovernance posture was even sharper when Barack Obama began the Sisyphean task of negotiating an international nuclear agreement with Iran. In the abstract, a deal appeared possible: Tehran wanted sanctions relief to improve its domestic economy; the international community wanted to prevent Iran from developing a nuclear weapons program. Both sides had an opportunity to reach their respective goals.

The odds of threading this diplomatic needle nevertheless appeared remote. The United States and Iran had barely spoken in a generation, ever since the end of the Iran hostage crisis in 1981, and the levels of mistrust between the West and Tehran couldn't have been higher. The task was made considerably more difficult by the fact that Iran's nuclear program had thrived during the Bush-Cheney era, expanding from under two hundred centrifuges to eight thousand.[27]

Obama nevertheless began the arduous process, applying harsh sanctions to bring Iran to the negotiating table and using existing alliances to assemble the five permanent members of the UN Security Council plus Germany.

GOP hawks had spent years arguing that diplomacy with Iran was folly, so if Obama's initiative proved to be a triumph, it would mean Republican orthodoxy on one of the world's most important national-security challenges had it backward.

Lacking confidence in their assumptions, and fearing the American president's success, GOP lawmakers took steps to ensure his failure. In January 2015, Republicans took the extraordinary step of partnering with a foreign government in the hopes of derailing an American diplomatic initiative. Lindsey Graham went so far as to travel to Jerusalem for a joint appearance with Israeli prime minister Benjamin Netanyahu, who also opposed diplomacy with Iran.

"I'm here to tell you, Mr. Prime Minister, that the [U.S.] Congress will follow your lead," Graham said, ignoring the fact that American elected officials don't traditionally take orders from foreign leaders.[28]

Soon after, House Speaker Boehner, circumventing the U.S. State Department, invited the Israeli leader to deliver an address to a joint session of Congress, as part of a not-so-subtle lobbying campaign intended to undermine the Obama administration's diplomatic efforts and weaken the American president's authority. As if the sensitive nuclear negotiations weren't challenging enough, U.S. officials also had to contend with the fact that elected Republican officials in their own country were waging a deliberate sabotage campaign against American foreign policy.

The GOP's partnership with a foreign country seemed to mark a new low, but the party had yet to hit bottom. In March 2015 forty-seven Senate Republicans, led by right-wing freshman senator Tom Cotton of Arkansas, wrote a joint open letter to Iran, warning the country not to trust the United States' word during the negotiations.[29]

To be sure, some GOP officials had dipped their toes in these waters throughout the Obama presidency. In June 2009, for example, Representative Mark Kirk traveled to China, where he reportedly told officials in Beijing not to trust the U.S. government's assurances on budget issues.[30] David Weidner, writing for the financial-news website MarketWatch, described the congressman's comments as "colossally stupid" and "dangerous."[31]

But the sabotage letter from forty-seven Senate Republicans was considerably more serious. The *Detroit Free Press* published a memorable editorial on the subject, arguing that the GOP letter "betrays a deep misunderstanding of our governmental structure, and a profound and dismaying disrespect for the office of the presidency, as well as its incumbent occupant. To disagree with a sitting president is one thing, even if that disagreement is loud, even if it is raucous. A deliberate attempt to undermine a sitting president's efforts to discharge his constitutional obligations is something else entirely."[32]

Adding insult to injury was the post-policy attitude undergirding the sabotage letter: its Republican signatories weren't scrambling to defeat an international agreement, because at that point in the process, there was no agreement. Rather, forty-seven GOP senators hoped to derail the *possibility* of a policy in the midst of delicate negotiations.

What's more, while American fights over foreign policy are as old as the country, the U.S. Senate Historian's Office told McClatchy in March 2015 that it could not find a comparable example in the chamber's history in which "one political party openly tried to deal with a foreign power against a presidential policy, as Republicans have attempted in their open letter to Iran."[33]

The circumstances bordered on farce. Start with two bitter enemies that hadn't even spoken in decades. Add heightened fears over terrorism and the most dangerous weapons ever created. Add five other major, disparate countries from around the globe, each with its own interests and agendas, and some of which fail to see eye to eye. Add delicate talks involving economics, nuclear science, diplomacy, national security, and geopolitics. Add the pressure that comes with knowing that failure would likely leave the world in an even more precarious position.

It was against this backdrop that one of the major political parties in the planet's dominant superpower abandoned the pretense of acting like a governing party and instead tried to frag its own country's foreign policy.

As the backlash against the GOP senators' stunt grew, some of the Republicans who signed the letter appeared to have second thoughts. Asked why he participated in an effort to undermine his own country's foreign policy, John McCain suggested he routinely put his name on documents without a lot of thought. "I saw the letter, I saw that it looked reasonable to me, and I signed it, that's all," he said. "I sign lots of letters."[34]

A day later, Rand Paul of Kentucky said he signed on as a way to "strengthen the president's hand" during the diplomatic talks.[35] The claim was every bit as incoherent as it seemed.

Others took pride in their handiwork. Marco Rubio's political action committee went so far as to send a fund-raising letter to donors, asking them to send contributions in order to reward the Floridian for endorsing the Iran letter.[36] The GOP senator added that he was so pleased with his party's message to Iran that he would have "sent another one tomorrow."[37]

Ultimately, the Republican campaign fell short. In July 2015 the United States and its coalition partners announced a breakthrough agreement—the Joint Comprehensive Plan of Action (JCPOA)—widely seen as one of the most unexpected diplomatic triumphs in a generation. The *New York Times*'s Max Fisher described the nuclear deal as "almost astoundingly favorable to the United States."[38]

The old-guard Republican establishment, including Colin Powell, Richard Lugar, and Brent Scowcroft, reached a similar conclusion and endorsed the policy. Former Virginia senator John Warner, onetime chairman of the Senate Armed Services Committee, also expressed public support for the agreement.

Predictably, most GOP officials engaged in over-the-top whining—Mark Kirk went so far as to say, "Neville Chamberlain got a better deal from Adolf Hitler"—but as was usually the case for the post-policy party, Republican arguments were long on hysterical rhetoric and short on substance.[39]

In mid-July 2015 Senator Lindsey Graham—at that point, a newly announced Republican presidential hopeful—appeared on MSNBC to denounce the international nuclear deal with Iran. Asked if he'd read the Joint Comprehensive Plan of Action, Graham conceded he had not.[40] GOP officials hated the deal, even if they didn't know its contents.

At its core, the GOP argument was rooted in a provocative contention: diplomacy with Iran is itself a misguided idea because there was no deal to make. Iranian officials should abandon their nuclear ambitions, Republicans argued, not in exchange for economic relief, but because American officials say so.

In October 2011 Jacob Heilbrunn contended in *Foreign Policy* that Americans were witnessing the "twilight of the wise man" and "the last gasp" of the Republican Party's foreign policy establishment.[41] Four years later, as the GOP railed against the Iran deal for reasons the party found difficult to explain, it seemed only fair to conclude the party's foreign policy establishment was done gasping and had completely lost its breath.

Left with limited options, many of these once prominent grown-ups in GOP foreign policy circles decided in 2016 that their party's ticket was simply undeserving of support. In June 2016 Richard Armitage, the deputy secretary of state in the Bush-Cheney administration and a longtime member of the Republican Party's foreign policy establishment, conceded, "If Donald Trump is the nominee, I would vote for Hillary Clinton."[42]

A week later, Brent Scowcroft, who served as White House national security adviser in two GOP administrations and led the Foreign Intelligence Advisory Board in the Bush-Cheney administration, also publicly threw his support to Clinton.[43]

Helping to drive home the point, Colin Powell also announced his backing of Clinton, and while George H. W. Bush and George W. Bush didn't endorse the Democratic ticket, they also withheld their support for the Republican ticket for the first time in their lives—a decision they were comfortable sharing with the electorate.

The more Donald Trump expressed hostility toward substantive foreign policy, the easier it was to understand the GOP old guard's anxiety. For example, the party's 2016 nominee spent months claiming superior judgment on matters related to foreign policy and national security by pointing to his 2003 opposition to the war in Iraq. Trump, however, was brazenly lying.[44] In September 2002, he appeared on a radio show and was asked directly whether he supported a U.S. invasion of Iraq. "Yeah, I guess so," Trump replied. At no time did he ever express any public opposition to the conflict.

In other instances, Trump seemed badly confused about current events. In a July 2016 interview with ABC News's George Stephanopoulos, the Republican nominee said emphatically that Russian President Vladimir Putin was "not going into Ukraine, okay, just so you understand. He's not gonna go into Ukraine, all right? You can mark it down. You can put it down. You can take it anywhere you want." When the host reminded the candidate that Russia had *already* entered part of Ukraine and claimed the territory as its own—two years before—Trump appeared lost.

This came on the heels of Trump's dismissing the Geneva Conventions, which serve as the basis for international humanitarian law, as "out of date,"[45] and endorsing the use of waterboarding and other methods of torturing suspected terrorists, even though intelligence experts asserted that they weren't effective for obtaining valuable information intelligence. At a 2015 campaign rally in Ohio, he told the crowd that even if such techniques did not work, "they deserve it anyway."[46] The height of post-policy thinking is politicians who embrace policies irrespective of their efficacy, and Trump was only too eager to embrace this approach without shame.

Presumably, the frontrunner for the Republican Party's presidential nomination—especially one with no experience in public service at any level—would be surrounded by a capable team of experts who could have advised and educated the candidate. But asked in March 2016 whom he spoke to on matters of international affairs, the future president replied, "I'm speaking with myself, number one, because I have a very good brain, and I've said a lot of things."[47]

A week later, Trump sat down with the editorial board of the *Washington Post* and starting naming a few of the foreign policy advisers who had his ear.[48] The short list included a twenty-eight-year-old low-level campaign staffer named George Papadopoulos, who wasn't a foreign policy expert and who was later sent to prison as part of his involvement in the scandal surrounding Russian interference in the 2016 elections. Another adviser cited by the candidate, Carter Page, wasn't a U.S. foreign policy expert, either, although he did draw the scrutiny of the FBI after describing himself in writing as an "advisor to the staff of the Kremlin."

In the transition period after Election Day, President-Elect Trump continued his wholesale indifference to his own country's foreign policies, starting with a December 2016 phone call with the president of Taiwan—a seemingly simple chat that broke with decades of precedent and sparked an international incident over the future of the "One China" policy, a delicate diplomatic position in which the United States formally treats Taiwan as part of China.

In a year-end White House press conference, Barack Obama was asked about the controversy and spoke at some length about the kind of general recommendations he'd made to his successor.[49] "What I've advised the president-elect," the Democrat said, "is that across the board on foreign policy, you want to make sure that you're doing it in a systematic, deliberate, intentional way. And since there's only one president at a time, my advice to him has been that before he starts having a lot of interactions with foreign governments other than the usual courtesy calls, that he should want to have his full team in place, that he should want his team to be fully briefed on what's gone on in the past and where the potential pitfalls may be, where the opportunities are, what we've learned from eight years of experience, so that as he's then maybe taking foreign policy in a new direction, he's got all the information to make good decisions."

After a brief recitation on U.S. policy toward Taiwan, under his and other modern administrations, Obama added, "That doesn't mean that you have to adhere to everything that's been done in the past. It does mean that you've got to think it through and have planned for potential reactions."

In other words, Obama recommended that the next president approach U.S. foreign policy in a way that reflected a degree of policy seriousness. His advice was less about what positions to take, and more about how to carefully and responsibly formulate those positions in the first place.

The result was a case study in post-policy politics: Obama suggested that Trump assemble a team, receive detailed briefings, assess recent history, and think through the consequences of dramatic changes in direction. Trump, after having heard this sound advice, thought it best to wing it and hope for the best. There was no strategy. There was no due diligence. There wasn't even a foreign policy to speak of. There was, however, an amateur who didn't know what he didn't know, and who lacked the interest and the wherewithal to learn.

In fact, not only did Trump's team choose not to coordinate with the U.S. State Department during the transition process, the president-elect decided he also didn't want to hear from his own country's intelligence community before taking office.

Immediately after the 2016 elections, Obama formally approved daily intelligence briefings for Trump and Mike Pence, which followed standard protocols: the incoming leaders would soon take their oaths of office, and they'd need to be up to speed to handle their responsibilities properly. Incoming presidents from both parties have traditionally sought out these briefings so that they're not caught off guard after taking office.

Trump, however, skipped nearly all of them.[50] The Republican made time for self-aggrandizing postelection campaign rallies, and he carved out hours to comment on *Saturday Night Live* sketches that hurt his feelings, but the president-elect didn't see the need to squeeze U.S. intelligence professionals into his schedule.

A month after Election Day, Trump sat down with Fox News's Chris Wallace, who asked about his disinterest in intelligence briefings ahead of Inauguration Day. The president-elect confirmed that the reports were true, and explained himself by saying, "I don't have to be told—you know, I'm, like, a smart person."[51]

A few weeks later, with his inauguration just days away, he described his ideal style of intelligence briefings: "I like bullets or I like as little as possible."[52]

In the months and years that followed, the U.S. intelligence community took a variety of steps to accommodate the amateur president's unique limitations and disregard for intelligence professionals' work. Just weeks into the Trump era, the staff of the National Security Council alerted the *New York Times* to the fact that while Obama preferred policy option papers that were three to six single-spaced pages, NSC staff was told in early 2017 "to keep papers to a single page, with lots of graphics and maps."[53]

NSC officials said later that they had resorted to strategically including Trump's name in as many paragraphs as possible because they found the president was more inclined to keep reading if he saw references to himself.[54]

Around the same time, *Foreign Policy* published a related report on U.S. allies in NATO (the North Atlantic Treaty Organization) "scrambling" to tailor an international gathering "to avoid taxing President Donald Trump's notoriously short attention span."[55]

"It's kind of ridiculous how they are preparing to deal with Trump," one source explained. "It's like they're preparing to deal with a child—someone with a short attention span and mood who has no knowledge of NATO, no interest in in-depth policy issues, nothing."

After White House National Security Adviser H. R. McMaster found that briefing the commander in chief on the ongoing war in Afghanistan was nearly impossible—Trump knew effectively nothing about the conflict and struggled to focus—a Trump confidant conceded, "I call the president the two-minute man. The president has patience for a half page."[56]

Reuters news service reported later that when U.S. officials wanted Trump to understand North Korea's nuclear capabilities, they brought him a model of North Korea's bomb test site with "a removable mountaintop and a miniature Statue of Liberty inside" to help him grasp scale.[57]

The problem did not improve over time. In April 2018 a former White House staffer said that when providing Trump information related to national security, the intelligence had to be reduced to "two or three points, with the syntactical complexity of 'See Jane run.'"[58]

But even those efforts to simplify fell flat. As 2018 came to a close, U.S. intelligence professionals complained to the *Washington*

Post that on most of the world's hotspots, Trump made decisions that were divorced from the judgment of his own country's agencies.[59] Warnings went unheeded, the classified President's Daily Briefing went unread, and the process in place to keep presidents informed unraveled. Even when briefings included "eye-catching graphics" intended to capture the chief executive's attention, the intelligence community routinely found "troubling gaps between the president's public statements and the facts laid out for him in daily briefings on world events."

In 2019, officials showed Trump a map of South Asia, and the president said he knew a couple of key details, identifying Nepal and Bhutan as being part of India.[60] It led to an awkward reminder that neither Nepal nor Bhutan is part of India. He also thought it was funny to mispronounce Nepal as "nipple" and refer to Bhutan as "button."[61] Noting the similarities between this and a scene from an episode of *The Simpsons* from 1995, in which Homer Simpson looks at a globe and finds the word "Uruguay" hilarious, *New York*'s Jonathan Chait joked, "It's like having Homer Simpson as president, but dumber."[62]

The impact on American governance was inevitable. The White House agenda was based almost entirely on Trump's hunches, parts of segments he'd seen on Fox News, and poorly thought out ideas that generated applause during assorted campaign rallies.

For example, in December 2015, candidate Trump had announced his support for a "total and complete shutdown of Muslims entering the United States." Like most of his policy pronouncements, the call wasn't the result of a detailed analysis or a review of what was feasible, legal, or just. The Republican was simply scratching an itch: Trump feared members of one of the world's largest faith traditions, so he decided the smart move would be to ban its adherents from American soil.

Trump's far-right audience roared with approval, which naturally led the candidate to assume his idea had merit. In fact, in the months that followed, he went further, raising the prospect of a government registry of American Muslims and suggesting he would "strongly consider shutting down mosques in the United States" if elected.

A week into his presidency, Trump issued an executive order clamping down on refugees and immigrants from select Muslim-majority countries from entering the United States. For those concerned with human decency, civil rights, religious liberty, and national security, it was an indefensible policy.

But Trump's Muslim ban was also crafted and implemented in the most post-policy way possible. White House officials started with the president's ridiculous campaign rhetoric and then worked backward, reverse engineering an executive order. The State Department had no input in the process.[63] The Department of Homeland Security[64] and the Justice Department[65] were isolated and excluded, too. Even many congressional Republicans were kept in the dark.

In fact, there was hardly a policy-making process at all. In an administration that valued the importance of sound governing, a decision like this would be scrutinized for months, vetted by a series of attorneys, and subjected to a lengthy bureaucratic review that examined, among other details, the practical effects of implementation. In January 2017 Trump and his team didn't bother with any of this—one can only conclude they didn't care.

The day after the president signed the order, airports around the world experienced the inevitable chaos that results from reckless policy making. Officials responsible for implementing Trump's policy were confused, and different facilities adopted varied and haphazard standards. A senior official at the Department of Homeland Security told NBC News, "Nobody has any idea what is going on."[66]

The same day, the president argued, "We were totally prepared. It's working out very nicely. You see it at the airports, you see it all over."[67]

Even by Trump standards, the boast bordered on delusion.

Not only were officials unsure how to implement the carelessly thrown together policy, they weren't sure whether it applied to green-card holders. There was a forty-eight-hour period in which the Department of Homeland Security, White House Chief of Staff Reince Priebus, and White House Chief Strategist Steve Bannon each gave contradictory answers on the proper status for green-card holders under the policy.[68]

A governing party should work out the fine points before rolling out a radical policy change, especially one this inflammatory and subject to scrutiny and opposition. This was the wrong policy, adopted for the wrong reasons, and implemented the wrong way.

Trump's approach to addressing the ISIS threat was no better. The Republican seemed to recognize the danger posed by the terrorist network, which was responsible for horrific attacks around the world and had ambitions of establishing a state-like caliphate in the Middle East. In 2015 Trump insisted that he had a secret plan to defeat ISIS.[69] By 2016, however, he pivoted and said he would ask U.S. military leaders to craft a plan for him.[70]

And in early 2017, the new Republican president settled on a policy to combat ISIS, which consisted entirely of "intensifying" the same policy Obama had already implemented and that was steadily driving the terrorist group out of Iraq without risking the lives of U.S. soldiers.[71] This became politically messy for the White House, since Trump had just spent a year and a half condemning Obama's policy toward ISIS. The Daily Beast website reported that White House officials asked Pentagon personnel to come up with ideas on how to help "brand" the Trump offensive against ISIS as "different from its predecessor."[72] In other words, Trump and his team didn't want to *look* as if they were continuing Obama's policy.

These absurd efforts, however, were carefully choreographed ballets compared with Trump's policy toward North Korea. The president began his term working from the assumption that he could simply get China to solve the problem of the rogue nuclear state for him.

China's president, Xi Jinping, traveled to Trump's Florida resort in 2017 and introduced his American counterpart to the details Trump should have learned during his eighteen months on the campaign trail. "After listening for ten minutes, I realized it's not so easy," the U.S. president said after his conversation with the Chinese leader.[73] "I felt pretty strongly that they had a tremendous power [over] North Korea. . . . But it's not what you would think."

In April 2017, as tensions between the United States and North Korea grew, Trump said he had a solution in mind: he'd dispatched a "very powerful" U.S. aircraft carrier to the Korean Peninsula as a show of force.[74]

It was never clear whether the president was lying or confused, but he hadn't done anything of the kind. The show of force existed only in his mind, for the ship in question, the U.S.S. *Carl Vinson,* was actually headed in the opposite direction. Once exposed, Trump's credibility on matters of national security suffered another avoidable setback, stemming from his inability to speak coherently and honestly.

Months later, Secretary of State Rex Tillerson seemed eager to ease tensions with Pyongyang. "We're trying to convey to the North Koreans, we are not your enemy," the top U.S. diplomat said at a State Department briefing. He added, "We would like to sit and have a dialogue with them about the future that will give them the security they seek and the future economic prosperity for North Korea."

Trump, apparently hoping to intimidate Supreme Leader Kim Jong-un into submission, announced soon after via Twitter that his administration's position was wrong. "The U.S. has been talking to North Korea, and paying them extortion money, for twenty-five years," the president wrote. "Talking is not the answer!"

Instead, Trump engaged in alliterative saber rattling—phrases such as "fire and fury" and "locked and loaded" sparked international headlines—that seemed to push the two countries closer to confrontation. At roughly the same time, he publicly praised Kim Jong-un's intelligence and said he'd be "honored" to meet the dictator.

In September 2017 the *Washington Post* reported that North Korean officials were quietly trying to arrange backchannel communications as part of an "attempt to make sense" of what exactly the White House's policy was.[75] No one could answer the question with confidence: the American president routinely contradicted himself and his team, and those looking for a clear U.S. foreign policy toward the dictatorship were left wanting.

In March 2018, on the heels of Trump saying there was no point "trying to negotiate with Little Rocket Man," the Republican welcomed South Korean officials to the White House. When one of them revealed that Kim Jong-un wanted to meet the American president for one-on-one talks, Trump agreed at once.

There was no policy review, no negotiations, no conversations with diplomats about the nature or scope of the bilateral meeting. In Trump's impulsive mind, it didn't matter that the United States was giving a nuclear-armed dictator exactly what he wanted—most notably the international legitimacy that comes from sharing the spotlight with the chief executive of the world's preeminent superpower—in exchange for nothing. There's no evidence Trump gave the matter deeper thought than this: it would be a dramatic and splashy photo-op.

The *New York Times* reported that events unfolded "haphazardly," which tends to happen when post-policy politicians pretend to govern without a plan.[76]

Two months later, Trump abruptly canceled the summit, fearing that North Korea might try to back out first. Republicans praised the move, saying Kim Jong-un wasn't trustworthy and shouldn't have been rewarded. Trump then impulsively reversed course again, restoring his plan to participate in the summit.

Making matters considerably worse, as the bilateral meeting drew closer, Trump refused to prepare. CNN reported that the U.S. leader "remained squarely focused on the summit's spectacle" and avoided "in-depth briefings about North Korea's nuclear program."[77] The commander in chief went so far as to publicly denounce the idea of doing substantive work ahead of the event.[78]

"I don't think I have to prepare," Trump told reporters ahead of the summit, adding, in reference to nuclear negotiations, "It's about attitude."

The results were predicable. The post-policy president, who agreed to the talks without any real thought or review, participated in nuclear negotiations without any preparation, and returned home empty-handed. North Korea pocketed the international legitimacy it wanted, and Pyongyang benefited further when Trump threw in the cancellation of U.S. military exercises with its ally South Korea. The United States got nothing.

It was possible that the summit could have produced a roadmap toward denuclearization, but it didn't. It was possible that Trump and Kim could have nailed down steps toward verifiability, but they didn't. It was possible that the two leaders could have reached an agreement on inspections, but they didn't. It was possible that North Korea could have agreed to a detailed accounting of its arsenal, but it didn't. It was possible that the two leaders could have established some kind of timetable for the near future, but they didn't.

The *New York Times*'s Nicholas Kristof explained, "It sure looks as if President Trump was hoodwinked. . . . Kim seems to have completely out negotiated Trump, and it's scary that Trump doesn't seem to realize this."[79]

On the contrary, Trump appeared convinced that he'd achieved a great triumph, assuring the world that he'd "solved" the problem posed by the dangerous dictatorship, to the point that North Korea was no longer a threat.[80]

"President Obama said that North Korea was our biggest and

most dangerous problem," Trump declared via Twitter in June 2018.[81] "No longer—sleep well tonight!" He went on to brag, "I signed an agreement where we get everything, everything."

Trump was either brazenly lying or hopelessly lost about the details of his own policy. It was as if he hadn't even read the vague and ultimately meaningless agreement he signed alongside Kim Jong-un, choosing instead to assume it would lead North Korea to give up its nuclear weapons.

A year later, in the wake of an even less productive second summit, the Trump administration had nothing to show for its efforts. North Korea welcomed the concessions the American president delivered, kept its nuclear weapons, and took steps to expand its nuclear program without penalty. Confronted with these awkward truths, Trump insisted in 2019 that his policy wasn't a total failure because North Korea had ceased its ballistic missile testing and was returning American remains from the Korean war. In reality, North Korea increased its ballistic missile testing[82] and stopped returning American remains from the Korean War.[83] Trump wasn't even prepared to defend his failure in a way that reflected an interest in governance.

The chaos continued. In March 2019 White House National Security Adviser John Bolton, Trump's third, and Treasury Secretary Steven Mnuchin touted new economic sanctions that the United States was poised to impose on North Korea in response to the rogue nuclear power's illicit shipping practices, intended to circumvent international sanctions. A day later, Trump issued a tweet announcing he was canceling those sanctions.

Several hours later, administration officials started telling reporters that Trump's tweet was actually a reference to some previously unknown sanctions, not the ones Bolton and Mnuchin had unveiled.[84] By that point, administration officials, blindsided and confused,[85] had no idea how, or even whether, to follow the instructions the president published on Twitter. [86]

Bloomberg News reported soon after that Trump indeed had

been referring to the sanctions announced by Bolton and Mnuchin, and officials had clumsily made up a false cover story in the hopes of making the president appear less bumbling.[87] Almost comically, in the end, White House officials managed to persuade Trump to leave his own administration's policy alone and not scrap the proposed sanctions. It was around this time that Richard Haass, a U.S. diplomat and president of the Council on Foreign Relations, a nonpartisan think tank, wrote, "There is not even the pretense of a national security process."[88]

Seemingly eager to prove Haass right, Trump visited the Korean Peninsula in June 2019, walked through the demilitarized zone separating North Korea and South Korea, and made a spectacle of a stroll on North Korean soil alongside his dictatorial pal. The *New York Times* reported a day later that in traditional U.S. administrations, "such a move might have been deliberated for weeks, put through an interagency process and approved only as part of a comprehensive approach to pressuring North Korea into giving up its nuclear program—a reward for progress."[89]

Of course, in normal U.S. administrations, decisions such as these are considered carefully as part of a national-security strategy, not what a confused amateur thinks might look cool on television. The made-for-TV spectacle gave Kim Jong-un a boost, serving up the kind of legitimizing moment his father and grandfather had craved but never achieved: hosting an American president in their isolated state.[90] North Korea's hereditary dictator had made the leader of the most powerful nation on earth travel around the globe to meet with him, and stand together as peers before the world's cameras. In exchange, Trump achieved nothing for his own country—except a fleeting photo-op, a simulacrum of diplomacy.[91]

Victor Cha, a Georgetown University professor and an expert on the Korean Peninsula, told MSNBC that Trump's meeting Kim on North Korean soil was like staging the Super Bowl at the start of the football season, before a single game has been played; a momentous meeting such as this would be expected as the culmination of a successful diplomatic process, not in lieu of diplomatic progress.[92]

Three months later, a reporter asked the American president whether he'd be willing to go to North Korea at Kim's invitation. "Probably not," Trump replied. "I don't think it's ready. I don't think we're ready for that. . . . I think we have a ways to go yet."[93]

The Republican's reluctance to go to North Korea might have made more sense if he hadn't already gone to North Korea.

Trump's approach to U.S. policy toward Iran was by most measures even more damaging. As a candidate, he spent a year and a half condemning the international nuclear agreement with Iran despite knowing very little about it. Trump repeatedly made up strange falsehoods about the policy—at one point, the future president claimed the deal allowed Iranian officials to "self-inspect"[94]—reinforcing impressions that he opposed it simply because the Obama administration had negotiated the policy.

Trump told voters, "My number one priority is to dismantle the disastrous deal with Iran."[95] After wrapping up the GOP nomination, he went so far as to say the deal is likely to "lead to nuclear holocaust."[96] But in early 2017, after taking office, Trump confronted a different kind of problem: the nuclear agreement with Iran was working exactly as intended. In fact, the administration formally notified Congress in April 2017 that Iran was living up to its commitments under the agreement.[97]

A few months later, the *New York Times* reported that the president had an hourlong meeting with Secretary of State Rex Tillerson, Defense Secretary James Mattis, White House National Security Adviser H. R. McMaster, and Chairman of the Joint Chiefs of Staff General Joseph Dunford, each of whom agreed on one thing: Trump should preserve the Joint Comprehensive Plan of Action.[98]

The president wanted his team to provide him with a pretense to abandon the policy he did not like for reasons he could not explain. When the top U.S. national security officials could not do that, Trump reportedly had "a bit of a meltdown."[99] It was an all too common pattern: when reality conflicted with his expectations, the president struggled to contain his fury toward reality.

In early 2018, U.S. allies from around the world launched a concerted and coordinated campaign with a narrow focus: convincing Trump that the JCPOA was working, so he had no reason to try to kill it. U.S. military leaders carefully sent the same signals, with army general Joseph Votel, the head of U.S. Central Command, telling the Senate Armed Services Committee that the Iran deal was both effective and important. "Right now, I think it is in our interest" to stay in the agreement, the general concluded.

The problem, of course, was that the pitch was predicated on the idea that the American president cared about the substance of the debate and would be swayed by appeals built around responsible governance. Those assumptions were badly misplaced. If Trump were serious about substantive details, he wouldn't have spent the previous three years condemning the policy he knew so little about based on nonsensical reasons.

European officials invested months of effort into negotiating with the administration, and, for a while, they thought their diplomatic campaign was having a constructive effect. U.S. allies held extensive negotiations with a State Department lawyer named Brian Hook, whose job was to help steer the existing policy in a more Trump-like direction. The result was a five-page document outlining supplemental additions to the Iran deal, intended specifically to satisfy the Republican president.

In the run-up to Trump's decision, French president Emmanuel Macron, German chancellor Angela Merkel, and British foreign secretary Boris Johnson each made separate visits to the White House to discuss the effective international agreement, and the *Washington Post* reported in May 2018 that each came away with "the feeling Trump had not read the five-page document they had prepared and perhaps was even unaware of the effort."[100] They made the mistake of bringing facts to a post-policy fight.

The diplomatic scramble to give Trump what he wanted and keep the existing framework in place fell apart, and the American leader turned his back on the deal anyway. When he told Macron about his final decision, the French president asked about the intensive work he and his colleagues had already done with Brian Hook, who led the negotiations on behalf of the U.S. government. Trump, ignorant and confused, replied, "Who is Brian Hook?"[101]

Summarizing the sentiment of the international community, CNN's Christiane Amanpour described the president's decision as "possibly the greatest deliberate act of self-harm and self-sabotage in geostrategic politics in the modern era."[102]

Trump didn't respond directly, though he did announce he was "ready, willing, and able" to negotiate a "new and lasting" nuclear agreement with Iran.[103] Of course, Iran already had done that, and Iranian officials held up their side of the bargain, only to see Trump cast it aside for reasons he could never explain coherently.

In June 2019 the president told reporters that if Iran was willing to forgo a nuclear weapons program, he would reward it financially. "They're going to be so happy," Trump said, imagining officials in Tehran taking him up on his offer to do what Obama had already done, creating economic incentives, including releasing frozen Iranian assets, for officials in Tehran to give up their nuclear ambitions.[104]

The underlying dynamic was as muddled as it seemed: Republicans, including Trump, had spent years arguing against providing Iran with economic incentives, insisting such an approach wouldn't work. When it did work, Trump killed the agreement anyway, only to turn around and offer Iran more economic incentives if it would only agree to do what it had already done.

GOP leaders made no effort to reconcile the contradiction, as if they simply didn't care enough to bother.

Elsewhere in the Middle East, the president's approach to policy making was equally vapid. The day before his 2017 inauguration, for example, Trump assigned his young son-in-law, Jared Kushner, to oversee the White House's efforts to reach an agreement between the Palestinian people and Israel. "If you can't produce peace in the Middle East, nobody can," Trump said to him at a pre-inaugural event. "All my life I've been hearing that's the toughest deal in the world to make. And I've seen it, but I have a feeling that Jared is going to do a great job."[105]

That "feeling" was based on nothing. At the time the president-elect made the comments, Kushner was a thirty-six-year-old aide with no experience in diplomacy or foreign policy, who would soon answer to a president with no interest in diplomacy or foreign policy. It was hardly a recipe for historic success.

Trump's confidence was nevertheless familiar: his indifference to policy complexities leads him to believe every problem is easy to solve, and he seemed eager to apply these unfortunate assumptions to the Middle East peace process. Three months into his term, the new president told Reuters, "There is no reason there's not peace between Israel and the Palestinians—none whatsoever."[106] As much of the world already knew, there were all kinds of reasons; Trump just didn't know enough to recognize them.

At the White House a month later, the president sat alongside President Mahmoud Abbas of the Palestinian Authority, the Palestinians' governing body, created by the 1993 Oslo Accords. Trump said that he'd heard how "tough" it would be to negotiate a Middle East peace agreement. "It's something, frankly, maybe not as difficult as people have thought over the years," Trump said, expressing confidence born of ignorance. He added there was a "very, very good chance" his administration would resolve the conflict.[107]

Those efforts were slow to advance, in part because Kushner was reluctant to familiarize himself with the details of the subject. In fact, the presidential son-in-law spoke to a group of congressional interns in July 2017 and expressed annoyance that others had encouraged him to learn relevant details before trying to negotiate a deal between Israelis and Palestinians.

"Everyone finds an issue, that 'You have to understand what they did then' and 'You have to understand that they did this.' But how does that help us get peace?" Kushner said. "Let's not focus on that. We don't want a history lesson. We've read enough books."[108]

In December 2017 Trump recognized Jerusalem as Israel's capital for reasons that appeared wholly detached from U.S. foreign policy and the peace process. Previous administrations of both parties had recognized Jerusalem's unique significance in negotiations between Israel and the Palestinians, and they saw little value in alienating the latter by taking such a step.

The *Washington Post* reported that senior White House officials quietly conceded, however, that Trump "did not seem to have a full understanding of the issue" and instead appeared focused on "seeming pro-Israel."[109] When administration officials raised foreign policy objections—both Rex Tillerson and James Mattis urged him not to do this—Trump didn't want to be bothered with substantive details.

"The decision wasn't driven by the peace process," one senior official said. "The decision was driven by his campaign promise."[110]

The circumstances were almost farcical. During the 2016 campaign, some of Trump's supporters told him they cared about the United States moving its embassy to Jerusalem and a diplomatic recognition of Jerusalem as Israel's capital. The then-candidate agreed to their appeals without any meaningful understanding of the position he was endorsing.

After winning the election, Trump started with a political posture: make his base happy, do what other modern presidents wouldn't do, "seem pro-Israel." From there he worked backward, instructing his staff to formulate a policy that would bolster his political calculation.

A *New York Times* report added that Trump "seemed to relish playing a familiar role: the political insurgent, defying foreign policy orthodoxy on behalf of the people who elected him."[111] In other words, explaining to Trump the importance of maintaining a delicate security balance, honored by both American parties and U.S. allies, didn't work—because he was the kid who enjoyed tearing down the class's tower of blocks, not the one who collaborated on building it.

Americans who thought they were electing a real estate developer to the presidency discovered they'd actually elevated a demolitions expert. Trump seemed to enjoy blowing up decades of carefully crafted, bipartisan foreign policy consensus for reasons he hadn't thought through.

Not surprisingly, Palestinian leaders were not impressed and walked away from the process, no longer able to trust the United States as a fair arbiter. In January 2018, while at the World Economic Forum in Davos, Switzerland, Trump tried to defend his move, explaining, "We took Jerusalem off the table. So we don't have to talk about it anymore."

The explanation made far less sense than he seemed to realize. As Slate's Joshua Keating explained, "I may not have written *The Art of the Deal,* but I'm pretty sure that this is not how negotiations work. If I were applying for a job and negotiating salary, benefits, and vacation days, then [was] told that I would be getting no vacation days at all so that the issue would be 'off the table,' I don't think this would make me more willing to compromise on salary and benefits."[112]

It wasn't long before Trump went out of his way to make peace even less likely, taking steps he made little effort to understand. In March 2019 the U.S. president announced recognition of the Golan Heights as sovereign Israeli land. It was a highly provocative move: Israel had seized the area from Syria by force decades earlier at the end of the Six-day War in 1967, and under international law, countries cannot keep territories taken after an invasion. It's why the United States had never recognized the area as Israeli land—in part to acknowledge the underlying legal principle and in part to discourage other leaders from claiming seized territories as their own in the future.

True to form, Trump arrived at his position impulsively and without any meaningful work on foreign policy. The American president was surprisingly candid on this point, boasting at an event in April 2019 about what convinced him to side with Israel: he was having an unrelated conversation with Kushner, U.S. ambassador to Israel David Friedman, and a lawyer named Jason Greenblatt, who used to work for the Trump Organization, when the subject of the Golan Heights came up.

"I said, 'Fellows, do me a favor. Give me a little history, quick. Want to go fast,'" Trump recalled. "'I got a lot of things I'm working on: China, North Korea. Give me a quickie.'"[113]

They gave the president a brief overview of the complex subject, at which point Trump proceeded to make a snap judgment on the spot to ignore decades of bipartisan American foreign policy, without regard for inconvenient details that he didn't find worthy of his time.

The president added, in reference to his administration's policymaking process, "We make fast decisions." The claim had the benefit of being true. What Trump did not say was that making decisions quickly is easy when those in positions of power pay no mind to public policy, history, substantive details, governing standards, or the practical implications of their decisions.

By the summer of 2019, the White House's approach to international affairs was unraveling. Trump's trade war against China was slowing the U.S. economy; his efforts to force leadership changes in Venezuela had failed; North Korea was conducting a series of ballistic missile tests despite Trump's effort to pretend otherwise; Iran was expanding its nuclear program; and the United

States' international standing was in sharp decline. It was around this same time that the American president expressed an interest in buying Greenland, which officials in Denmark treated as a deeply unfortunate joke.

In the months that followed, the pattern continued, as Trump inexplicably abandoned our Kurdish allies in Syria and escalated tensions with Iran in a series of moves that seemed to defy reason.

The *Washington Post*'s Anne Applebaum summarized the cause of the breakdown: "There is no plan; there is only whim and instinct. There is no international orientation either, just a series of hasty unplanned, unexamined decisions, followed by Twitter tantrums."[114]

Around the time Applebaum's column was published, Trump administration officials were engaged in diplomatic talks with Taliban leaders in the hopes of ending the U.S. war in Afghanistan. A tenuous agreement took shape: American troops would withdraw, and the Taliban would provide counterterrorism assurances. By September 2019, the basic structure of a deal appeared to be in place.

That is, until the post-policy president had a few ideas. Excited by the prospect of a diplomatic spectacle, Trump invited Taliban leaders and Afghan president Ashraf Ghani for peace talks to be held at Camp David—the presidential retreat nestled in the wooded hills of Maryland, about an hour northwest of Washington—around the time of the anniversary of the 9/11 attacks. The *New York Times* reported, "After staying out of the details of what has been a delicate effort in a complicated region, Mr. Trump wanted to be the deal maker who would put the final parts together himself, or at least be perceived to be."[115]

In a familiar dynamic, said the *Times,* the White House's plan "was put together on the spur of the moment." There was no due diligence, no discernable process, and none of the usual scrutiny from the professionals on the National Security Council. Predictably, the process collapsed soon after, in part because Trump involved Afghani officials in the capital of Kabul whom the Taliban considered illegitimate.

Democratic congressman Ruben Gallego of Arizona, a retired U.S. Marine who served in Iraq, asked a reasonable question: "Who the fuck thought it was a good idea to invite the Taliban to Camp David—let alone around September 11th?"[116] It was, of course, the American president himself, who had an unexamined whim and hadn't bothered to think through any of this.

When a reporter asked Trump about how he arrived at his decisions, he replied, "Actually, in terms of advisers, I took my own advice. I liked the idea of meeting. . . . I didn't even—I didn't discuss it with anybody else."[117] That he considered this something to boast about reinforced his indifference to a deliberate and responsible policy-making process. A few days later, the *New Yorker*'s Susan Glasser concluded, "The intensive national-security decision-making process of previous presidents, Republican and Democrat alike, has been abandoned by Trump, subverted to the presidential ego, and will not return for the duration of his tenure."[118]

CHAPTER 6

"The Cruelty Is the Point"

The Collapse of Immigration Policy

IMMIGRATION REFORM WAS ONCE a unique issue on the American political landscape. While most modern partisan disputes walk along entrenched battle lines concerning the size of the government, use of taxpayer funds, federal borrowing, and the reach of public insurance programs, since the Bush-Cheney era, the immigration debate has unfolded largely outside the usual spectrum.

For nearly two decades, Republican officials have said their priority is increased border security, while Democratic officials have sought legal protections for undocumented immigrants who are already in the United States. No one needs an army of policy wonks to recognize the obvious fact that the respective goals have not been in conflict. The parties could strike a compromise that satisfied both sides.

The political benefits of such a breakthrough would be significant, but the broader policy stakes are every bit as important, if not more so. There are roughly eleven million undocumented immigrants in the United States, many living in the shadows, making them a vulnerable national community in need of legal recognition. Their future carries economic, moral, and security consequences that reverberate in every part of the country.

What's more, given the degree to which immigration has been central to the country's fabric and traditions, it's difficult to deny the importance of writing the next chapter in the American story. Few are satisfied with the status quo, creating obvious demand for a bipartisan compromise.

In 2006 the Senate passed just such a bill, which was co-authored by Senators John McCain, the Republican from Arizona, and Massachusetts Democrat Ted Kennedy. The legislation's provisions featured familiar elements: increased border security, a guest worker program, and a pathway to citizenship for undocumented immigrants who pass a background check, pay a fine, and learn English.[1] It enjoyed the support of the Bush-Cheney White House, Republican lawmakers such as Kentucky's Mitch McConnell, and Democrats—including a young senator from Illinois named Barack Obama.

But it failed anyway when far-right House Republicans said they couldn't accept the kind of concessions Democrats included in the bill. GOP leaders knew the measure had the votes to pass, but they honored a fabricated standard called the Hastert rule, associated with Dennis Hastert, the longest-serving Republican Speaker of the House in history: it's not enough for a proposal to enjoy majority support; bills won't even come up for a vote unless a majority of Republicans back it. (There is no literal rule; it's an optional procedural tool party leaders have utilized—before and after Hastert's tenure—for the sake of political convenience.)

The House GOP's inflexible position was driven by a fear of the party's far-right base, which expected Republicans to reject anything conservatives could characterize as "amnesty"—a vague term that tended to refer to any policy that allowed undocumented immigrants to remain in the United States. Many of the party's members came to believe a vote for a bipartisan trade-off would invite an immediate primary challenge and brutal coverage on Fox News; consequently, much of the House Republicans' rank and file found it far easier to reject any plan that included provisions Democrats liked.

It was a sign of things to come for a GOP that proceeded to abandon its status as a governing party, especially on the issue of immigration.

Four years later, the Obama administration and congressional Democrats tried again to pass a comprehensive reform package built on the framework of the McCain-Kennedy bill, working from the assumption that if some Republicans were comfortable with the bill endorsed by George W. Bush, especially in the Senate, they'd consider legislation that was effectively identical.

Those assumptions were misplaced. In 2010, Democrats controlled both chambers of Congress, but they needed some GOP support to reach the sixty-vote threshold in the Senate. Republicans balked, telling the Obama White House to focus on border security first, and then they'd think about bipartisan legislating.

The Democratic president, eager to solve the problem, proceeded in good faith. In fact, Obama eagerly touted this fact in May 2011 during a speech in El Paso, Texas, where he reminded observers that his administration had strengthened border security, deployed unmanned aerial drones to patrol the border, and tripled the number of intelligence analysts working at the border. The president wondered aloud whether Republicans were prepared to take yes for an answer.

Reflecting on the fact that pleasing GOP officials seemed increasingly impossible, the president joked sardonically, "Maybe they'll need a moat. Maybe they want alligators in the moat. They'll never be satisfied. . . . The question is whether those in Congress who previously walked away in the name of enforcement are now ready to come back to the table and finish the work we've started."[2]

The answer to that question was no. Throughout Obama's

first term, as the border grew more secure, Republican interest in working on the issue evaporated.

The political calculus seemed to change, at least for some in the party, at the midpoint of the Obama era. Mitt Romney, the Republican Party's presidential nominee in 2012, ran on a platform that opposed a mass roundup of undocumented immigrants, but also rejected a pathway to citizenship. Asked what these immigrants already in the United States were supposed to do in this limbo, the GOP candidate pointed to an ugly, two-word solution: "self-deportation."

The idea, which the Republican nominee was occasionally forced to discuss grudgingly, was to make life in the United States for undocumented immigrants so miserable, they'd voluntarily deport themselves from American soil.

This did not prove to be an electoral winner, but Romney's tactic failed so spectacularly that it briefly altered the trajectory of the debate. In 2004 John Kerry's Democratic ticket with Senator John Edwards of North Carolina had received 53 percent of the Hispanic vote. Four years later, Barack Obama's Democratic ticket with Delaware senator Joe Biden earned 67 percent of the Hispanic vote. But in 2012, with Mr. Self-Deportation on the ballot, Obama's share of the Hispanic vote reached 71 percent.[3]

Republicans couldn't help but take note of the results. The incentive to deny the Democratic president a major legislative victory was gone—Obama couldn't seek a third term—while the incentive to take the issue off the table with a bipartisan bill was strong. Suddenly the door was open.

Just forty-eight hours after Obama celebrated his reelection win, House Speaker John Boehner told ABC News, in reference to immigration, "This issue has been around far too long. A comprehensive approach is long overdue, and I'm confident that the president, myself, others can find the common ground to take care of this issue once and for all."[4] The same day, Mark McKinnon, a former top aide to George W. Bush and John McCain, added, "The best thing about Republicans losing is that it will likely force them to cut an immigration deal."[5]

Less than a week before Obama's second inaugural, Senator Marco Rubio, whose Cuban-born parents immigrated to the United States in 1956 and who has long expressed a keen interest in the issue, sketched out his vision for an immigration package with the *Wall Street Journal,*[6] to the delight of the Democratic White House: the Florida senator's plan read like it had been largely copied and pasted from Obama's 2012 blueprint.[7]

Two weeks later, John McCain effectively conceded that the president had held up his end of the bargain with enforcement efforts during his first term in the White House. "There has been real improvements in border security," the GOP senator said.[8] When a reporter asked about the impact this would have on passing a bipartisan immigration bill, McCain added, "I think it helps a lot."

In mid-April 2013 the so-called Gang of Eight—four Senate Democrats and four Senate Republicans—unveiled a comprehensive immigration blueprint that wasn't dissimilar to the bipartisan bill from 2006. Writing for the *New Yorker,* Ryan Lizza reported that GOP lawmakers even privately lobbied Fox News on-air hosts on the bill in the hopes that the network wouldn't direct other congressional Republicans to kill it.[9]

Proponents of the bill got an additional boost in June 2013 when the nonpartisan Congressional Budget Office reported that the legislation would have the effect of adding millions of additional taxpayers, which in turn would reduce the nation's budget deficit by hundreds of billions of dollars.[10] It was around this time that the bill received endorsements from business leaders, labor unions, prominent law enforcement organizations, many leaders from the faith community, and the nation's leading immigrant advocacy groups. Independent polling showed the American mainstream embraced the bill, too.

With all of these pieces in place, it seemed GOP opponents were going to have a hard time finding reasons to reject the compromise package. Yet they did anyway: although the measure passed the Senate in June 2013, thirty-two of the chamber's forty-six Republicans—about 70 percent of the GOP conference—voted to kill the bill.

The measure then went to the GOP-led House, where John Boehner initially assured the public that he and his leadership team would "let the House work its will."[11] But the Speaker did not even try to honor his commitment. A month after making the comments, and nine months after insisting that "a comprehensive approach is long overdue," Boehner reversed course and said the House would not consider the bipartisan compromise—even if most of the chamber was inclined to vote for it. The Republican leader added that he no longer supported a "comprehensive" solution.

House Republicans had the option of writing and passing their own immigration bill, which would lay the groundwork for a bicameral compromise, and, for a while, that appeared to be a distinct possibility. "The House is going to do its own job in developing an immigration bill," Boehner said in July 2013. He added, "It is time for Congress to act. But I believe the House has its job to do, and we will do our job."[12]

Almost immediately, the post-policy party decided it had no intention of doing its job. Republican representatives understood they faced a problem in need of a solution, and they knew there was a sensible compromise on the table, but they still weren't prepared to govern. And so a new political effort got under way: GOP officials went in search of an excuse to justify their refusal to do meaningful work on one of the nation's most urgent and consequential issues. Some Republicans made the case that the brutal civil war in Syria demanded lawmakers' attention, which meant immigration reform would have to wait.[13] Others in the

GOP, echoing a foolish talking point from the health care debate, claimed the Senate compromise bill had too many pages.[14]

Other Republicans said the party needed to focus all of its political energies on criticizing the Affordable Care Act, and addressing the immigration issue would change the subject from the controversial health care law.[15] This argument had the benefit of candor, but it was nevertheless underwhelming; it left no doubt that GOP officials were more interested in waging political attacks than solving a problem.

Eventually the party seized on one talking point: Republican lawmakers weren't prepared to pass any major legislation on any issue because they saw Barack Obama as a lawless and untrustworthy tyrant who would refuse to follow federal statutes as written.

"There's widespread doubt about whether this administration can be trusted to enforce our laws," Boehner contended in February 2014.[16] Eric Cantor[17] and Paul Ryan[18] used nearly identical language to rationalize their party's disinterest in legislating.

Even by the standards of contemporary politics, it was weak tea, obviously intended to shift the blame for policy failure away from the very people who refused to govern. But it was made worse when Senator Chuck Schumer of New York, a member of the Democratic Party's leadership, called the GOP leaders' bluff by announcing that Democrats were prepared to delay implementing the immigration compromise package until 2017, when Obama would no longer be president.[19]

Reform advocates were generally on board with this approach, knowing that it would take a long while, even after the legislation passed, to write the relevant federal regulations needed for full implementation.[20] Yet, once again, even after their most prominent objection had been addressed, House Republicans rejected the compromise offer anyway. They simply weren't prepared to do any meaningful work. Partisans may have bickered over who ultimately bore responsibility, but in this game of Clue, there was no doubt about the identity of the killer: it was Republicans, in the House, with their indifference.

Ahead of the 2014 midterm elections, Marco Rubio, one of the principle architects of the Senate compromise, backed away from his own legislation.[21] John McCain did the same.[22] The National Republican Senatorial Committee ran attack ads ahead of Election Day condemning supporters of the immigration reform plan that *four Republican senators had coauthored*.[23]

Soon after the election votes were tallied, and GOP officials gained new power, Barack Obama decided he had no choice but to try to address parts of the problem though executive actions. Congressional Republicans implored him not to act unilaterally, insisting that White House actions would preclude bipartisan legislating on the issue.

The humor behind the pitch was lost on them: GOP lawmakers who refused to work on immigration policy were, in effect, threatening the president by telling him they'd refuse to work on immigration policy.

Two weeks after the 2014 midterms, the Democratic president took advantage of his executive powers, announced a new federal policy extending new protections to millions of undocumented immigrants. In his remarks announcing the policy, Obama added, "And to those members of Congress who question my authority to make our immigration system work better, or question the wisdom of me acting where Congress has failed, I have one answer: pass a bill."

In response to the White House announcement, GOP officials turned to their stockpiles of fury, insults, and red-hot disgust. What the party did not have was a public policy or anything resembling a serious, substantive approach to the issue at hand. As Republicans prepared to take control of both chambers of Congress for the first time in nearly a decade, Vox's Ezra Klein explained in November 2014, "The Republican policy on immigration needs to be something more than opposing Obama's immigration policies.

It needs to be something more than vague noises about border security. . . . Democrats support a path to citizenship. Republicans don't support anything."[24]

Ahead of the 2016 campaign, the Obama administration believed it had a compelling success story to tell about immigration. As the *Washington Post* reported, "As the Department of Homeland Security continues to pour money into border security, evidence is emerging that illegal immigration flows have fallen to their lowest level in at least two decades."[25] R. Gil Kerlikowske, Obama's commissioner of U.S. Customs and Border Protection, added, "We have seen tremendous progress. The border is much more secure than in times past." Paradoxically, Obama had achieved the Republicans' border security dreams in spite of Republicans' best efforts to stop him.

In true post-policy fashion, GOP presidential hopefuls, however, spent months railing against the Democratic president for all but ignoring border security. Even among this field of White House hopefuls, Donald Trump stood out, basing much of his candidacy on the promise that he'd get Mexico to finance the construction of a giant concrete wall that covered much of the U.S.-Mexico border, from the Pacific Ocean to the Gulf of Mexico.

As Joshua Green reported in his 2017 book, *Devil's Bargain: Steve Bannon, Donald Trump, and the Storming of the Presidency,* the impetus from the Republican's campaign team was to use the idea of a wall as a mnemonic device: Trump needed help remembering the broader issue as part of his standard stump speech.[26] Sam Nunberg, a top adviser to the candidate, explained that it was a memory trick that clicked. "Roger Stone and I came up with the idea of 'the wall,' and we talked to Steve [Bannon] about it," Nunberg said. "It was to make sure he talked about immigration."

And that he did. Trump conceded to the *New York Times* in April 2016 that he saw the rhetoric as catnip to be sprinkled during his far-right rallies. "You know, if it gets a little boring, if I see people starting to sort of, maybe thinking about leaving, I can sort of tell the audience, I just say, 'We will build the wall!' and they go nuts," the candidate said.[27]

There was some truth to that. If "Yes We Can" was the optimistic mantra that helped propel Barack Obama's candidacy in 2008, "Build the Wall" was similarly embraced by Trump's followers as motivating fuel.[28] It would have been far more compelling, however, if the presidential hopeful had any idea what he was talking about. Trump argued in April 2016 that, if elected, he would force Mexico to make a onetime payment to the United States of between $5 billion and $10 billion, and he'd be able to compel America's neighbor to write the check by threatening to cut off remittances: money transfers Mexican immigrants routinely make to help loved ones back home. Ending the payments would do enormous damage to the Mexican economy, so Trump assumed he could use this as leverage to pry wall money from officials in Mexico City.

As was often the case, the Republican hadn't thought this through. The United States generally wasn't in the habit of launching extortion schemes against allies and key trading partners. Moreover, even if Mexico paid Trump's ransom, $10 billion wouldn't come close to funding a border wall. But perhaps most importantly, legal experts concluded that Trump's scheme would almost certainly run afoul of existing federal laws.[29]

The future president would later deny ever having proposed such an idea, his published campaign platform notwithstanding.

It was all part of a post-policy posture that Trump made little effort to hide. Asked about his unorthodox ideas, the candidate told Bloomberg News six months before Election Day, "I'm not sure I got there through deep analysis." He added that he sometimes learned things from random supporters he'd meet while signing autographs at rallies.[30]

When the interview turned specifically to immigration, Trump conceded that he opposed the Gang of Eight reform plan without knowing what it was and came to believe the U.S.-Mexico border was a mess based on what he knew instinctively.

It was an unusual admission from the man who'd just locked up the GOP presidential nomination. Not only is it impossible to instinctively know substantive border policy details, but the political fight surrounding the Gang of Eight's immigration bill was a politically dominant issue for the better part of a year. Given Trump's stated interest in immigration and the Mexican border, it stood to reason that a prospective presidential hopeful would at least familiarize himself with the basics of the debate.

But he didn't, choosing instead to launch a presidential campaign predicated in part on his immigration views, which consisted of a few offensive sound bites—none of which was the result of "deep analysis."

In August 2016 Trump sat down with Fox News's Bill O'Reilly, who asked the candidate to clarify his vision of immigration policy. The Republican proceeded to praise existing federal immigration laws as "very strong" (a position he'd later abandon), said the Obama administration had already deported "tremendous numbers" of criminals (another position Trump would later abandon), and added that, if elected, he'd simply enforce existing laws "perhaps with a lot more energy" (a position that was largely indecipherable).

What was painfully clear was that Trump didn't know or care about public policy—even pertaining to one of his signature issues, upon which he based much of his candidacy. He couldn't answer questions about basic substantive details because he'd evidently given them very little thought.

Many hoped that after taking office, Trump would have no choice but to shake his post-policy habits and at least try to govern like an actual president. Those hopes were soon dashed.

Early in his term, the Republican headlined a campaign rally in Iowa where he declared support for what he characterized as a fresh proposal: newly arrived immigrants "should not use welfare for a period of at least five years." When his supporters responded with a rapturous standing ovation, Trump added, "We'll be putting in legislation to that effect very shortly."

As it turned out, this wasn't necessary: Bill Clinton signed the same policy into law in 1996.[31] Trump didn't know that, hadn't bothered to check, announced a plan without a moment's worth of due diligence, and passed along an old idea as if he'd just come up with it.

On multiple occasions, the president's ignorance was as much a hindrance as an embarrassment. In 2018, for example, it was the Trump administration's position that Congress needed to approve funding to hire new immigration judges to help clear more border cases. It was also in 2018 that the president denounced the hiring of new immigration judges, in large part because he didn't know who they were or what they did.

At an event in South Carolina, Trump described a purported conversation with his team on the subject: "They come up, and this was an order, this was—'Sir, we need five thousand judges.' I said, 'Five thousand?' So, we put a judge on, like, on the bench, federal, it takes us weeks to vet, it takes us a long time to get the judges, one—we're talking about one person. And they want five thousand, I said, 'Where are you going to find five thousand people to be judges? How many do we have now?' 'I don't know the number.' They don't even know the number, even though they're in charge, okay? Nobody knows the number. We have thousands of judges already."

Nearly all of the harangue was gibberish. No one, for example, had recommended hiring five thousand immigration judges; the entire country had fewer than four hundred immigration judges at the time. The White House had recommended funding for hundreds more.[32]

But more importantly, as part of his denunciation of his own administration's position, it never occurred to Trump to learn the differences between Article III federal judges, who are confirmed by the Senate and serve lifetime positions as part of an independent branch of government, and immigration judges, who are part of the Justice Department. He simply heard the word "judge" and had a knee-jerk reaction. Learning the basics of the policy and making a decision accordingly—the way mature officials who care about governing behave—never seemed to occur to him.

A year later, in an apparent reference to the immigration judges his administration wanted to hire, Trump told reporters, "To be honest with you, you have to get rid of judges."[33] Everyone involved had no choice but to simply ignore the president's confusion. To get rid of immigration judges would have wreaked havoc on the existing system, while undermining due process. Trump treated these consequences as annoying trivialities he didn't need to understand.

This happened more than it should have. Throughout his presidency, the Republican was firm in his insistence that a giant border wall—a nonissue for the GOP before Trump's rise to power—was necessary to protect the nation's interests, but when he tried to make a coherent case in support of the medieval vanity project, his reasoning fell apart.

Trump argued, for example, that a wall would dramatically curtail illegal immigration, which was hard to take seriously, since most undocumented immigrants in the United States are here on overextended visas. He was similarly convinced that a wall would end the illicit drug trade, despite the evidence that smugglers routinely try to sneak drugs through ports of entry.

The more the president was reminded of these inconvenient details, the more he'd reject the evidence that conflicted with his assumptions. "Don't believe people when they say it all comes through the portals; it doesn't—the ports of entry," Trump assured the public, referring to drug trafficking. "It comes through—the big loads come through the border, where you don't have wall, where they can drive a truck, a big truck, loaded up with drugs."[34]

Trump's own Drug Enforcement Administration, among others, published data that directly contradicted these claims.[35] A president interested in competent governance might have reevaluated false assumptions based on the evidence, but Trump condemned the figures he found objectionable and moved on. His principal arguments in support of his signature idea were based almost entirely on a faulty grasp of the most basic elements of the debate.

At one point, the chief executive traveled to Texas for an event in which he examined photographs of tunnels criminals had dug to smuggle guns and drugs across the border. A report in *Time* picked up on the obvious point that the White House apparently missed: "Neither border patrol agents nor President Trump explained how a border wall would help stop the flow of drugs through tunnels."[36] The same presentation featured money seized from a suspected criminal who had overstayed a visa—which was another problem a wall couldn't address.

Trump also insisted that a wall would combat human trafficking, and to bolster his point, the president spent months describing horrific scenes of people being transported into the United States, bound and gagged in the back of vehicles. U.S. Border Patrol officials scrambled to substantiate Trump's stories with real-world evidence.[37] They came up empty.

The White House nevertheless pretended not only that his border wall had real merit, but also that there was a vital plan in place that administration officials were implementing with pride. In the president's first year in office, the Trump administration welcomed bids for border wall prototypes, to be built in Southern California. Taking advantage of information collected through a Freedom of Information Act request, *USA Today* uncovered in October 2017 an unusually "haphazard" system in which there was "continuous confusion over the most basic details of the process—deadlines, page counts, how to submit bids, where to submit bids, and so on."[38] It was another example of GOP officials struggling to achieve their own ends because they didn't know enough about how government works.

The newspaper spoke to the head of a consulting firm for government contractors who said, "This shows me the government still does not know what they want, is still developing specifications, and spending a great deal of money trying to figure out what they want to do."

The assessment was unambiguously true: Trump knew he wanted something he could call a "wall," checking a box on his list of ill-considered campaign promises, but neither he nor his team had a meaningful understanding of any of the other relevant details. When the president argued via Twitter that his plan "never changed or evolved from the first day I conceived of it,"[39] he was brazenly lying: Trump spent years changing practically every facet of his idea, from its height to its length, from its cost to the materials that would be used in its construction.[40]

Indeed, the president even changed his mind about the prototypes. After an in-person presidential inspection and months of boasts about the structures in the California dirt, Trump quietly scrapped the plan and decided he actually wanted steel slats along the border. In an apt metaphor, jackhammers unceremoniously destroyed the structures, reducing them to dust in the wind, after the White House decided it no longer cared about them.[41] Trump had spent millions of taxpayer dollars to finance the prototypes' construction, and they turned out to be little more than expensive and meaningless props for a political photo-op.

As 2018 came to a close, and policy makers faced a deadline to prevent a government shutdown, Trump made border wall funding his sole concern, issuing an unsubtle ultimatum to lawmakers: he would agree to keep federal operations going only if Congress approved taxpayer funding for his border barriers.

As the winter holidays approached, and the political world was gripped by the very real possibility that there would be no eleventh-hour solution, the president published on Twitter a never-before-seen image of black steel bars. "A design of our Steel Slat Barrier which is totally effective while at the same time beautiful!" Trump wrote on December 21, 2018.[42]

No one outside the Oval Office knew what to make of it. For one thing, the image was a bit of a mess: the enlarged close-up of the pointed tips of the steel bars bore little resemblance to the rest of the image. For another, even the White House's congressional allies had no idea why Trump had rolled out a new, unexamined image. "The president put out a tweet of a picture with spikes on top of fencing; that's not even in the conversation," said Senator James Lankford, a Republican from Oklahoma. "That's not even in one of the designs the Border Patrol has proposed."[43]

Trump, winging it, didn't care, and since there was no governing blueprint to follow, the president and his team had no qualms making up stuff as they went along. Acting White House Chief of Staff Mick Mulvaney announced two days later that the president's previous plan, which the Republican had touted for two years, had been replaced, though administration officials hadn't bothered to tell anyone. "The president tweeted out a picture yesterday [of] the steel fence, the steel slatted fence with a pointed top and so forth, that's what we want to build," Mulvaney said.[44]

It wasn't long before Trump was micromanaging the steel-slats initiative, giving his team specific directives. A *Washington Post* report painted an almost comical picture of an obsessed president barking instructions about the color of the barriers.[45] And the shape of their tips. And their height. And what they should be called. And the number of gates. And the size of the gates. And the construction schedule. And "the minutiae of contracts."

The *Post* article added that the president's frenzied oversight led to "frequently shifting instructions and suggestions" that in turn "left engineers and aides confused."

Responsible governing from mature policy makers this was not. By the White House's estimation, there was a complex crisis unfolding at the border, raising questions related to national security, diplomacy, international laws, and humanitarian interests. It would have been difficult for even competent presidents to navigate the logistical and legal challenges, but there was Trump, focusing his attention on paint colors and the shape of bollards.

All the while, the president invested enormous energies into telling the public that his administration had already successfully expanded a physical barrier along the U.S.-Mexico border, and that rascally fact-checkers who said otherwise were not to be believed. The Republican even instructed his followers at various rallies to stop chanting "Build the wall!" and to start chanting "Finish the wall!" Implicit in the directions was the idea that Trump was well on his way toward completing his goal.

This is precisely why the White House was so displeased when the *Washington Examiner,* a conservative outlet, reported in July 2019 that the administration had not installed "a single mile of new wall in a previously fenceless part of the U.S.-Mexico border in the 30 months since President Trump assumed office."[46]

The *Examiner* added that U.S. Customs and Border Protection, which was responsible for overseeing border barrier construction, "confirmed that all the fencing completed since Trump took office is 'in place of dilapidated designs' because the existing fence was in need of replacement."[47]

The truth was straightforward: old barriers had, in some cases, been replaced, but the percentage of the border in which there was no structure separating the two countries hadn't changed since Trump took office. The U.S.-Mexico border is 1,954 miles long, and as the Obama era came to an end, there were 654 miles of physical barriers along the border.[48] Nearly three years later, that number had not changed, despite the president's incessant claims to the contrary.

Because of these confounding details, Trump found it difficult to make up his mind about how best to talk about his own record. On the one hand, the president insisted that the United States' southern border was wide open, and a wall was necessary to end the scourge created by undocumented immigrants pouring into the country.

On the other hand, Trump also wanted the public to believe the opposite. In August 2017, for example, he bragged, "The border was a tremendous problem and [Homeland Security officials are] close to eighty percent stoppage. Even the president of Mexico called me—they said their southern border, very few people are coming because they know they're not going to get through our border, which is the ultimate compliment."

Officials conceded soon after that there was no such discussion with President Enrique Peña Nieto of Mexico. (Trump has long had an unfortunate habit of describing conversations that only occurred in his mind.[49]) The 80 percent figure, too, was made up out of whole cloth.[50] Far from a one-off misstep, the incident was part of a larger pattern in which Trump took credit for progress, while simultaneously arguing that there'd been no progress.

Over the course of a single day in December 2018, the president tweeted in the morning, using his idiosyncratic approach to capitalization, "Our Southern Border is now Secure and will remain that way."[51] That same afternoon, he added, "Drugs are pouring into our country. People with tremendous medical difficulty and medical problems are pouring in, and in many cases it's contagious. They're pouring into our country. We have to have border security. We have to have a wall as part of border security."[52]

Unable to choose which narrative he liked better, Trump pushed both, indifferent to the contradiction.

To be sure, there have been daunting challenges at the border for many years, but none would be substantively addressed by building giant concrete slabs or installing giant steel slats, which is probably why practically no one from either party pushed the idea before Trump took office. Nevertheless, if the president had been serious about the endeavor, he could have done what most other presidents have done for generations: try to make a deal with the lawmakers who have the power of the purse, offering them concessions in exchange for achieving his priority.

Trump didn't know how to do that; aides in the West Wing didn't even trust him to try to negotiate competently with Democrats who might exploit his ignorance. And so a more radical approach, untethered to real governing, took shape.

Trump's first maneuver was to target a group of young immigrants, known as Dreamers, who entered the country illegally as children, usually accompanied by their families. The group got its name from the Dream (Development, Relief and Education for Alien Minors) Act: a bipartisan compromise bill crafted in 2001 to help these immigrants become American citizens. As the GOP shifted to the hard right, the party abandoned the legislation, deeming it a Democratic priority.

The Obama administration protected Dreamers though a policy called Deferred Action for Childhood Arrivals (DACA), which shielded them from deportation and allowed them to work and study in the United States through temporary permits.

As a candidate, Trump promised to end DACA and deport the kids, though, as with most issues, he gave no indication of having given the matter much thought.[53] As a president-elect, the Republican's line softened, saying, "We're going to work something out that's going to make people happy and proud. . . . They got brought here at a very young age, they've worked here, they've gone to school here. Some were good students. Some have wonderful jobs."[54]

And as president, Trump said in April 2017 that the Dreamers could "rest easy" about his agenda, telling the Associated Press that he was "not after the Dreamers, we are after the criminals." He added, "That is our policy."[55]

Five months later, Trump abandoned that policy and rescinded Dreamers' legal protections. For these hundreds of thousands of young people for whom the United States has been the only home they've ever known, the administration's announcement was a nightmare come to life. They were already part of the American fabric—from soldiers to students, workers to homeowners—who were suddenly confronted with the threat of deportation for reasons the president lacked the ability to explain.

But on top of the cruelty were the widespread concerns that the president had no meaningful understanding of the implications of his own decision. The *New York Times* reported, "As late as one hour before the decision was to be announced, administration officials privately expressed concern that Mr. Trump might not fully grasp the details of the steps he was about to take."[56]

Nevertheless, the White House created a six-month window in which the president expected lawmakers to come up with a legislative solution to the mess he'd just created after having told Dreamers they could "rest easy."

A week later, Congress's top two Democrats—Chuck Schumer and Nancy Pelosi of California, the minority leaders of their respective conferences—both of whom were desperate to help Dreamers, announced that they'd reached an agreement with the president, effectively trading DACA protections for increased border security. Trump largely confirmed the contours of the deal, adding that the fight over the border wall would "come later."[57] The president soon after backed away from his deal without explanation.

In January 2018, as the process moved forward, Trump held a televised meeting in the White House Cabinet Room with lawmakers from both parties and both chambers. The peek behind the curtain was informative. In one especially memorable moment, viewers saw Trump accidentally endorse a Senate Democrat's request for a clean DACA bill extending protections to Dreamers with no strings attached. It seemed like a breakthrough—right up until someone reminded the president that this wasn't his position.[58]

At the same gathering, Trump was unexpectedly candid, telling lawmakers they should work out an immigration plan and send it to his desk for a signature, and he'd keep his distance from the process because he didn't much care what was in the bill.

"I will say, when this group comes back, hopefully with an agreement, this group and others from the Senate, from the House, comes back with an agreement, I'm signing it," the president said, leaving himself little wiggle room. "I mean, I will be signing it. I'm not going to say, 'Oh, gee, I want this or I want that.' I'll be signing it." Trump went on to say that if lawmakers negotiated an immigration package "with things that I'm not in love with," he intended to embrace it anyway, regardless of the substantive details.

Reform advocates briefly found his indifference encouraging because it made the prospects for a solution appear more likely. But the blue skies soon darkened: Trump's blithe posture wasn't part of an actual plan or strategy, and it wasn't long before the president, untethered to a policy agenda, was soon persuaded to do the opposite of what he'd promised.

A couple of days after saying he'd sign whatever members came up with, Trump rejected another bipartisan DACA compromise that had been reached by Senators Lindsey Graham and Dick Durbin.[59]

Durbin and Trump spoke over the phone on January 11, and the Illinois Democrat was under the impression that the blueprint he and Graham had sketched out would get the president's approval. Trump invited the two senators to the White House to wrap up the agreement.

They arrived to discover that right-wing opponents of the bipartisan deal—including Representative Bob Goodlatte of Virginia and Arizona senator Tom Cotton—had also been invited to the meeting. Stephen Miller, a thirty-three-year-old far-right presidential adviser, had scrambled behind the scenes to bring

Republican hard-liners into the mix because he feared that more mainstream lawmakers would take advantage of the president's unfamiliarity with the basics of immigration policy.

Miller and his allies considered it likely that Durbin and Graham would "trick" Trump into endorsing an insufficiently conservative plan, since the post-policy president wouldn't be able to tell the difference between proposals he was supposed to like and those he was expected to oppose.[60]

A few weeks later, Senators John McCain and Chris Coons, a Democrat from Delaware, unveiled another bipartisan immigration compromise.[61] Trump rejected it, too.[62]

It was at roughly this point that White House demands took a turn that was equally aggressive and confusing. Trump expected a DACA deal that would end "chain migration"—a process through which American citizens bring loved ones to the United States through family-sponsored immigration—though the president routinely talked about the issue in a way that suggested he didn't fully understand what it was.[63] Trump also railed against the Diversity Immigrant Visa lottery, though the president's rhetoric suggested the details of that program were also lost on him.[64]

Ultimately, Trump told Congress that he also expected significant cuts to legal immigration—a position once limited to the fringes of far-right politics, and one that differed from his positions before[65] and after[66] the DACA fight—or there could be no deal. It was not a constructive approach to putting out the fire the president himself had started. In all, lawmakers presented the White House with six different bipartisan plans, each of which Trump rejected,[67] despite his earlier assertions.

From the perspective of the president—or at least those who were helping guide his hand—he had all the leverage he needed. Trump could package his unpopular ideas with Dreamer protections that enjoyed broad support, confident in the knowledge that he could prevail either way. If Congress paid the ransom, Trump would get his win. If lawmakers balked, he'd punish Dreamers, whom the Republican Party's base didn't like anyway.

In February 2018 many congressional Democrats were even prepared to finance Trump's border wall—to the tune of $25 billion—in exchange for a twelve-year path to citizenship for Dreamers. The White House said it wasn't good enough. Asked on Fox News about his plans for a wall, the president said, "The Democrats have actually agreed to that. But I have to get more."[68]

By the time the Senate took up competing immigration alternatives, they all failed: Trump wouldn't support the bipartisan solutions, and Democrats wouldn't support the far-right alternatives. The president's preferred approach needed sixty votes in the Republican-led Senate. It received thirty-nine.[69]

The author of (the ghostwritten) *The Art of the Deal* couldn't muster the will or the skill to work out a governing solution to the mess he'd created. For all intents and purposes, he didn't even try.

Trump proceeded to pretend that Democrats didn't really care about the Dreamers' fate, because if they did, they would have embraced a far-right agenda that built a wall and slashed legal immigration. There was no mystery behind the cynicism: the president intended to sacrifice his hostages and blame those who failed to pay his ransom.

As part of the same strategy, Trump implored Senate GOP leaders to end the legislative filibuster altogether so that his immigration preferences could pass, seemingly unaware of the fact that the sixty-vote threshold was irrelevant in this instance, given the woeful support for his preferred far-right plan.

After noting that the president didn't seem to understand how Congress works, the *Washington Post*'s James Hohmann explained, "There is no strategy. There is no message discipline. There is no process." There is only a president who "does not think through the second- and third-order consequences of his decisions."[70]

Trump didn't realize judges would step in and let the hostages go, but that's exactly what happened: as the legislative process collapsed, the federal courts ruled that Trump couldn't scrap the DACA program.[71] Suddenly his leverage was gone.

Lawmakers threw the president a hanging curve, right over the center of the plate: a bipartisan deal that would have given him $25 billion for a border wall and popular DACA protections for Dreamers. If he'd accepted it, this would have been a high point for his entire presidency and provided him something to brag about ahead of his 2020 reelection campaign.

Trump didn't know or care enough about governing to swing at the ball.

Reeling from his border wall funding fight, in the spring of 2018, Trump adopted a zero-tolerance approach toward undocumented families trying to enter the United States, which included separating children from their parents. This included asylum seekers, who had a legal right to seek protection in the United States.

"We have to break up families," Trump said at a White House event. "The Democrats gave us that law. It's a horrible thing. We have to break up families. . . . It's terrible."[72]

The practice was, in fact, horrific. People in the United States and around the world were mortified by images of isolated children, having been forcibly separated from their families, being housed liked prisoners by the United States government.

The Department of Health and Human Services's Office of Inspector General visited roughly half of the facilities that housed migrant children and found kids who would "cry inconsolably," according to one of the office's reports.[73] A notorious facility in Texas, accused of not providing children with adequate food or access to soap, was described by lawyers as "appalling."[74]

But when the president blamed this tragedy on "that law," he was pointing to a statute that did not exist in reality. There was a 2008 bill signed by George W. Bush, intended to address child trafficking, which required U.S. officials to release child migrants traveling alone in the "least restrictive setting" while their cases were adjudicated. Generally, that meant putting the kids in the care of family members or official shelters.

It did not require children to be separated from their parents, which is why neither Democratic nor Republican officials took such steps prior to 2018. Trump and his team did it anyway, indifferent to the human costs and abandonment of human decency.

The White House struggled to defend its breathtakingly cruel policy, swerving wildly from one nonsensical talking point to another. The president said he had no choice under the Bush-era law, which wasn't true. He then said the tactics were necessary to create a disincentive to other would-be migrants, but that was wrongheaded, too. A July 2018 study, published by the Center for American Progress and written by political scientist Tom K. Wong of the University of California, San Diego, found that family separation and detention policies did not deter migrants from trying to reach the United States.[75]

Trump even went so far as to say his policy was the Obama administration's fault, which was delusional and insulting to the public's intelligence. The administration had the option of simply taking responsibility for its own approach to governing, but that would have required a degree of integrity the president and his team lacked.

As the *Washington Post*'s Catherine Rampell explained, "The Trump administration's goal is to inflict pain upon these families. Cruelty is not an unfortunate, unintended consequence of White House immigration policy; it is the objective."[76] Or as *The Atlantic*'s Adam Serwer famously put it, "The cruelty is the point."[77]

But as with so many Trump administration policies, the president and his team didn't have the technocratic wherewithal to even monstrously mistreat families in a coherent way. Before the White House settled on its zero-tolerance policy in April 2018, officials hadn't bothered to coordinate in advance with U.S. Immigration and Customs Enforcement, the Department of Health and Human Services, U.S. Citizenship and Immigration Services, or even Border Patrol officials.[78]

An investigation by the nonpartisan Government Accountability Office later scrutinized the consequences of the administration's incompetence and found that relevant agencies "did not take specific planning steps" to implement Trump's callous policy because no one told them it was coming.[79] The White House, careless about the functions of federal power, effectively blindsided its own team.

Exacerbating the crisis, the Trump administration began separating the families without crafting any kind of coherent bureaucratic monitoring system. "It's hard to imagine, isn't it?" Democratic senator Dick Durbin said in a congressional hearing.[80] "The United States of America says, 'We're going to start separating kids forcibly from their parents. We're not going to set up a system of tracking those kids so we can find them or match them back up with their parents. We're just going to separate them.' . . . There were some very basic ways that we could have kept track of the children and parents."

Those basic ways were ignored, as were warnings of the looming disaster.[81]

By June 2018, the administration was so rattled by the scope of its failure that it could barely speak coherently on the subject. Over the span of a few days, Trump said his policy was an awful development that should be blamed on Democrats;[82] Attorney General Jeff Sessions said it was a great policy justified by the Bible;[83] First Lady Melania Trump said she wanted "both sides" to fix her husband's policy;[84] and Homeland Security Secretary Kirstjen Nielsen thought it would be a good idea to publish a tweet that read, "We do not have a policy of separating families at the border."[85]

By one count, the president and top administration officials, clearly making stuff up as they went along, changed their story about the purpose and the wisdom of the family separation policy fourteen times.[86]

Two months into the zero-tolerance crisis, Axios reported that the president viewed the family separation issue "as leverage": he'd consider rolling back the policy, but only in exchange for concessions such as border wall funding.[87]

Trump wasn't especially coy about his thinking, connecting his family separation policy to his immigration demands. "We could have an immigration bill," the Republican said at a mid-June White House event.[88] "We could have—child separation—we're stuck with these horrible laws. They're horrible laws. What's happening is so sad—is so sad. And it can be taken care of quickly, beautifully, and we'll have safety. This could really be something very special. . . . We can do this very quickly if the Democrats come to the table."

It reinforced the impression that Trump saw desperate families and scared migrant children as bargaining chips. The fact that Democrats had already been at the negotiating table, offering the president bipartisan immigration compromises that he carelessly dismissed, was a detail Trump chose to bury. (The same week Trump made these comments, the *New York Times* reported on negotiations among House Republicans on an immigration bill intended to make the White House happy.[89] Congressional Democrats, who were eager to participate in the legislative talks, were deliberately "cut out of negotiations.")

In practice, the president had effectively swapped one group of young hostages for another: if he couldn't get a border wall by threatening Dreamers, maybe he could get one by taking children from their families and putting them in cages.

When his ransom again went unpaid, Trump, facing an international backlash and brutal news coverage, signed an executive order on June 20, 2018, officially ending his zero-tolerance policy.

The move came just five days after the president insisted the policy couldn't legally be addressed through an executive order.

But even at that point, the Trump administration's approach to governance was so incompetent that no one was sure what the presidential order meant. The Department of Homeland Security had been left in the dark about relevant details,[90] and different agencies within the administration were under the impression that Trump's order meant entirely different things.[91]

Politico reported that relevant departments were "gripped by confusion,"[92] while the *Washington Post* added, "The slapdash nature of the effort was apparent when the White House released an initial version of the executive order that misspelled the word 'separation.'"[93]

Vox's Matt Yglesias further explained that Trump's entire approach reminded Americans that the president "does not know anything about public policy, diplomacy, constitutional law, or legislative strategy. So you get instead what he's delivered over the past two weeks—aggressive hostage-taking, lying, trolling, chaos, dissembling, and cruelty."[94]

Six months later, as 2018 drew to a close, the president looked anew for fresh leverage, launching the longest government shutdown in American history, undermining the economy and leaving many unpaid federal workers to turn to food banks to get by.[95]

Trump told congressional Democrats his shutdown would last indefinitely until they rewarded him with billions of dollars in wall funding. Trump's strategy had a childlike simplicity: stripped of his other hostages, the president would shut down the government, lawmakers would write a big check, and the White House would claim victory.

Even as the strategy failed, at no point did it occur to Trump to act like a president. If the Republican genuinely believed the nation would benefit from a giant wall, Trump had the option of crafting a persuasive argument, taking his pitch to the people and their representatives, and negotiating in good faith. It's the standard model in American Politics 101.

It's also a model with which Trump was wholly unfamiliar. The Republican never saw the value in learning about the policy, taking the details seriously, or governing like an adult in a position of authority in a global superpower. Indeed, after his shutdown extortion gambit collapsed, the president issued an emergency declaration giving himself emergency powers to spend public funds on border barriers in defiance of Congress's wishes.

While insisting he had no choice but to circumvent his own country's legislative appropriations process—expressing particular alarm about a so-called caravan of Latin American migrants, which served as the basis for a variety of strange conspiracy theories—Trump told reporters, "I didn't need to do this, but I'd rather do it much faster. . . . I just want to get it done faster, that's all."[96] The comments left little doubt that the president himself realized there was no actual emergency that required an unprecedented power grab. This was a matter of convenience for an ostensible leader who couldn't negotiate deals or work within a legitimate policymaking framework.

In the months that followed, Trump used the power he'd given himself—the authority to redirect federal funds without lawmakers' permission—to raid the Pentagon budget, taking funds allocated for military construction. This included, in several instances, the construction of schools and day care facilities for the children of American service members, and sending those funds to pay for steel border slats.[97]

It led Andrew Exum, a deputy assistant secretary of defense in the Obama administration, to note on Twitter, "I once repurposed some excess funds we had earmarked for Ebola-related contingencies to buy blankets and cooking fuel for refugees fleeing the Islamic State and had to get four separate congressional committees to sign off on it before I could spend a dollar."[98] In August 2019, however, Trump took billions of dollars from the U.S. military—not excess funds, but money that had already been assigned to specific priorities—without any congressional approval, all in service of a fantasy emergency.

It was one of many instances in which the president put aside good governance and instead concocted a series of dubious schemes related to immigration, ranging from ridiculous to inhumane.

In March 2019 Trump made an unscripted comment during a campaign rally in Michigan, telling supporters he was prepared to "close the damn border" altogether unless Mexico took steps to make him happy. "Very simple," the president said. "It's all very simple. People try and make it too complicated. It's all very simple."

For those who are unfamiliar with the substance of governing, everything is always "very simple."

According to a report in *Time,* after the crowd erupted in applause, Trump was "electrified" and "told aides he wanted to move ahead with a plan to close ports of entry."[99] Such a drastic move would have sweeping economic consequences in both countries, but the president chose not to think along those lines: a far-right audience approved of his off-the-cuff remark, which was all the analysis he needed.

To prevent the border closing, Trump added on Twitter, Mexico would have to "immediately stop ALL illegal immigration coming into the United States throug [*sic*] our Southern Border."[100] A reporter asked the next day if he was concerned about his plan adversely affecting the U.S. economy. "Sure, it's going to have a negative impact on the economy," the president replied, but he intended to do it anyway.

Trump ultimately retreated from his threat, crediting increased security measures from Mexico. Officials in Mexico had no idea what the American president was talking about—they'd done nothing differently—because Trump had really just looked for a face-saving way to back down from the bluff he hadn't thought through in the first place.

A month later, White House officials thought it would be a good idea to round up detained immigrants and transport them to so-called sanctuary cities—a conservative buzz phrase used to describe progressive communities that resist federal efforts to identify and remove undocumented immigrants.[101] Under this strategy, the administration would simply drop off detainees on these cities' streets, apparently as a way of retaliating against the president's domestic political enemies.

The initial reports on this said the Department of Homeland Security promptly shot down the nonsense, and an official statement from the White House said, "This was just a suggestion that was floated and rejected, which ended any further discussion."[102] But in the Trump administration, where the right hand routinely did not know what the even-further-to-the-right hand was doing, the president soon after turned to Twitter to say the opposite, telling the world that he was "giving strong considerations to placing Illegal Immigrants in Sanctuary Cities only."[103]

Even after administration officials said the plan was illegal and impractical,[104] Trump said the opposite.[105] The president didn't even seem to notice that by broadcasting consideration of the absurd maneuver, he was inadvertently making worse the problem he ostensibly wanted to improve: the Republican was publicly talking up a policy in which his administration would transport new arrivals—for free—to diverse and welcoming American cities with large immigrant communities.

Max Boot, a Republican pundit before the Trump era, wrote a column explaining, "This idea isn't half baked. It's not even a quarter baked. That Trump is even talking about this shows the White House's intellectual bankruptcy on the president's No. 1 political issue."[106]

A few weeks later, Trump boasted at a campaign rally that he'd already begun delivering undocumented immigrants to progressive communities, despite the fact that he hadn't, and indifferent to the fact that he couldn't take such a step under his own country's laws.[107] As far as the Republican president was concerned, however, legal limits belong in the same category as policy details: they were pesky niceties that got in the way and were better left ignored.

In his first address to a joint session of Congress the month after his inauguration, Trump told Americans, "We must restore integrity and the rule of law at our borders." It proved to be one of the more ironic remarks of his presidency.

After Trump forced Kirstjen Nielsen from her post as Homeland Security secretary in April 2019, the *New York Times* reported that the president had been in a habit of calling Nielsen at home "to demand that she take action to stop migrants from entering the country, including doing things that were clearly illegal, like blocking all migrants from seeking asylum. She repeatedly noted the limitations imposed on her department by federal laws, court settlements, and international obligations."[108]

Her responses, the article added, "only infuriated" Trump.

Soon after, CNN reported that the president traveled to Southern California, where he told border agents to ignore federal laws and deny entry to all migrants. "If judges give you trouble, say, 'Sorry, judge, I can't do it. We don't have the room,'" Trump instructed them.[109]

Officials later sought guidance from their superiors, asking whether to follow the presidential directive to ignore the legal limits under which they're supposed to operate. (They were told to ignore Trump's rhetoric.) A senior administration official added, "At the end of the day, the president refuses to understand that the Department of Homeland Security is constrained by the laws."[110]

CNN reported further that Trump told the commissioner of Customs and Border Protection, Kevin McAleenan, ahead of his six-month stint as acting secretary of the Department of Homeland

Security, that if he faced legal repercussions for ignoring American laws, Trump would reward him with a presidential pardon.[111]

The same week, the *New York Times* reported that Trump had empowered presidential adviser Stephen Miller, who's been accused of promoting theories popular with white nationalist groups, to oversee immigration policy in the White House.[112] The article described a series of incidents in which Miller, ostensibly speaking on behalf of Trump, demanded administration officials do more to deny welfare benefits to legal immigrants, work around court-ordered protections for migrant children, and make the review process more difficult for those seeking asylum.

Some of Trump's own political appointees—not career officials—pushed back against policies they considered "legally questionable, impractical, unethical, or unreasonable." According to the *Times,* this in turn "further infuriated" those in the White House who didn't care for the existing federal laws and preferred to circumvent them. At a meeting in which officials at Citizenship and Immigration Services were on the receiving end of a Miller tirade, they believed "it was almost as if Mr. Miller wanted asylum officers to ignore the law."

In some cases, the ambiguities disappeared. The *Washington Post* reported in August 2019 that Trump was feeling so desperate to expand border barriers ahead of his reelection campaign that he'd directed aides to "aggressively seize private land and disregard environmental rules."[113] Those who were caught running afoul of the law, Trump added, would also be rewarded with presidential pardons.

Confronted with questions about contracting procedures and the proper use of eminent domain, the president reportedly replied that he expected administration officials to simply "take the land." After nearly three years in office, Trump was so hostile toward American policy making that the lines between governing and corrupt authoritarianism had blurred to the point that he no longer bothered to notice them.

CHAPTER 7

"We Stand by the Numbers"

The Federal Budget

FOR MUCH OF ITS history, the Republican Party was sincerely concerned about the nation's budget deficits and national debt. Fiscal responsibility was, by any fair measure, a staple of GOP orthodoxy—not as a matter of political convenience, but as a deeply held principle.

In a contemporary context, by contrast, the party's traditional positions on fiscal issues are unrecognizable. Following World War II, Republicans were broadly supportive of extremely high top marginal tax rates, arguing that the taxes were necessary in order to help pay off war-era debts and finance postwar reconstruction. Throughout Dwight Eisenhower's presidency, the top marginal rate was more than 90 percent.

During John F. Kennedy's presidency, the United States had a small debt and almost no deficit, so the Democratic president was comfortable cutting taxes in 1963—with the wealthiest American households receiving the fewest benefits—as part of a policy the White House described as "peace dividends."[1] Many Republicans of the era balked: GOP lawmakers feared the country simply couldn't afford Kennedy's tax breaks.[2] The party was fulfilling its role as a fierce anti-deficit party. It would be decades before Republicans became a fierce anti-tax party.

The motivations behind the shift are complex, but they were driven in part by a conservative backlash to Lyndon Johnson's Great Society agenda—most notably its goals of ending domestic poverty—and perceptions on the right about "liberal" excesses and the "welfare state." The result was a dramatic shift in the GOP's entire posture, starting in earnest during Ronald Reagan's tenure, when Republican tax cuts pushed the deficit to depths without modern precedent.

George W. Bush's presidency was even more extraordinary: after inheriting enormous surpluses from the Clinton administration, the Republican president and his GOP allies put two tax breaks, two wars, a Medicare prescription drug plan to benefit seniors, and a Wall Street bailout on the nation's credit card. David Walker, the nation's comptroller general under both Clinton and Bush, from 1998 to 2008, described the latter as "the most fiscally irresponsible president in the history of the republic."[3]

Between the financial crisis of 2008–09, rising unemployment, domestic industries teetering on the brink, and a global downturn, the economic conditions that Barack Obama inherited couldn't have been more dire. And yet, upon taking office, the Democrat from Illinois faced immediate demands to balance the budget from the same Republicans who'd uprooted the nation's fiscal foundations. The GOP's overwhelming hypocrisy and cynicism were obvious, but what was less appreciated was the degree to which the party was generally incoherent on the subject.

In the first year of the Obama era, for example, many Republican leaders with indefensible fiscal records struggled to explain their metamorphoses. Senator Orrin Hatch, for example, told the Associated Press that "it was standard practice not to pay for things" during the Bush-Cheney era.[4]

It was a curious observation. According to the Utahan—at

the time, in the thirty-second year of his forty-two-year career on Capitol Hill—GOP policy makers, en masse, effectively suffered from mass amnesia for nearly a decade about one of the party's core priorities. As a Democratic president furiously tried to rescue the country from the grips of the Great Recession, and it made economic sense to authorize sweeping public investments without regard for their fiscal impact, Hatch and his brethren decided it was time for an entirely new "standard practice"—which was the polar opposite of the one they'd just embraced.

Maine's Senator Olympia Snowe added, by way of a defense of her party, "Dredging up history is not the way to move forward."

In *Monty Python and the Holy Grail,* there's a scene in which Sir Lancelot, mistakenly convinced he needs to rescue a damsel in distress, storms into a castle during a wedding party and indiscriminately slaughters guests with his sword. The castle owner, eager to curry favor with the powerful knight, immediately urges the survivors to let bygones be bygones.

"Let's not bicker and argue about who killed whom," he tells his few remaining guests.

Snowe's perspective was annoyingly similar. In effect, she didn't want to see the political world bicker and argue about which party destroyed which country's finances.

Around this same time, congressional Republicans, under pressure from Democrats to stop whining and start engaging, decided to unveil a budget plan designed to rival the blueprint introduced by the Obama White House. A top House GOP official told Politico, "We need to hold something up and say, 'Here are our charts. Here are our graphs. It's real.'"[5]

It must have seemed like a good idea at the time. In practice, Republican leaders presented a deeply strange document with no numbers and no budget estimates. The party's unusually vague plan said it intended to reduce the deficit, but it didn't specify by how much, when, how, or why. Instead, the purported Republican plan featured assorted buzz phrases written inside randomly chosen shapes, connected through a series of lines that didn't appear to serve any real purpose.

As an art project, it was incomprehensible. As a governing document, it was pitiful.

TPM reported that by the time Republicans had finished premiering their alleged budget, reporters had one big question: "Where's the *actual budget*? You know, the numbers that show deficit projections and discretionary spending?"[6]

With the release of the bizarre document, House Republicans became the first major party in recent memory to unveil a budget plan that omitted a budget. The GOP had grown sensitive to criticisms that it had devolved into a party of no, so it threw together a plan that left Republicans vulnerable to entirely different attacks: they'd become the party of no numbers.

The supposed point was to give the public the impression that Republican policy makers were members of a governing party, when, in fact, GOP leaders were so detached from serious policy making that they ended up proving the opposite.

Hari Sevugan, in his capacity as the Democratic National Committee's press secretary, was only too pleased to take advantage of the Republicans' ineptitude, issuing a statement that read, "After 27 days, the best House Republicans could come up with is a 19-page pamphlet that does not include a single real budget proposal or estimate. There are more numbers in my last sentence than there are in the entire House GOP 'budget.'"

In the months that followed, Republican officials remained obsessed with deficit reduction, despite their own troubled record on the issue and the desperate need for policy makers to prioritize stronger economic growth as the nation slowly tried to recover from the Great Recession.

However, the GOP's aim was not to promote a better-balanced budget. Rather, nearly all the party's focus was on cutting federal expenditures. When Republicans were pressed for specifics, John

McCain tried to answer the call, publishing a top-ten list of what he described as the most egregious examples of wasteful spending. Compiling the list, he added, was "a lot of fun."[7]

It may have been entertaining, but it wasn't nearly as substantive as the senator seemed to think. For example, McCain pointed to a $650,000 federal investment in "beaver management" in North Carolina and Mississippi. "How does one manage a beaver?" the Republican asked.[8] The truth, which he neglected to explore, was that beaver management was designed to help water distribution affecting a series of farms and infrastructure projects in the Southeast. By investing $650,000, wildlife agencies were able to save $5 million in prevented flood damage.

"Maybe you should ask him how much he knows about this and why he picked it out for ridicule," Congressman David Price, a Democrat from North Carolina, said of McCain. "We know why he chose this—because it sounds funny."[9]

McCain was also among a group of senators who crafted a bill to create a bipartisan commission on deficit reduction, which would be responsible for pushing policy makers to make uncomfortable decisions about the nation's fiscal future. GOP leaders told the White House that if Obama publicly endorsed the bill to create the commission, it stood a better chance of passing in the Democratic-led Congress. The president agreed and threw his support behind the measure.

At this point, curiously, several Senate Republicans who'd cosponsored the bill decided to reverse course and vote against their own proposal. According to them, the commission idea had to die because its panelists might have recommended a compromise combining spending cuts and increased government revenue in order to bring the budget closer to balance.[10] It's also likely that Republicans defeated the bill they'd helped write for the simple reason that the Democratic president had endorsed the legislation.

Either way, the incident was a reminder of how petty the party had become and of how selective it was about applying its purported fiscal principles.

There are basically only two ways to reduce the federal budget: the government can take in more revenue and/or it can spend less. Republicans made clear that they were prepared to categorically rule out one of them.

Sure, they said, the deficits the GOP had helped run up tore at the fabric of society. And sure, they claimed, balancing the budget was absolutely necessary to rescue future generations from hardship and ruin. In March 2009 Republican governor Mark Sanford of South Carolina said he was so concerned about the deficit that he saw accepting funding from the Democrats' Recovery Act as tantamount to "fiscal child abuse."[11]

But those same GOP policy makers also said they would not consider any proposal that raised any tax on anyone, at any time, by any amount, for any reason. The parties were nevertheless supposed to work cooperatively and in good faith on the budget and address the nation's fiscal challenges.

To no one's surprise, constructive policy making in this area proved impossible. In July 2011, as a debt ceiling deadline loomed, Barack Obama and Vice President Joe Biden hosted a meeting with the top eight lawmakers in Congress—four from each party and each chamber—to gauge their preferences for a debt reduction target: $2 trillion, $3 trillion, or $4 trillion. A majority of the participants, including House Speaker John Boehner, eyed the most ambitious goal.

At the same meeting, however, Eric Cantor and Jon Kyl balked.[12] From the far-right lawmakers' perspective, there was no point in even trying to reach such a target, since a bipartisan agreement would necessarily ask Republicans to make at least some concessions. Cantor and Kyl, speaking on behalf of their party's rank-and-file members, made clear that in their vision of governing, there could be no give-and-take between the parties. GOP officials were prepared to accept drastic cuts to public investments and social insurance programs—and nothing else.

Soon after, to help resolve the Republican-imposed debt ceiling crisis, the parties agreed to form a supercommittee whose members were tasked with crafting a multitrillion-dollar debt reduction plan. It failed spectacularly when GOP lawmakers on the panel said they'd consider only a solution in which Democrats accepted concessions and Republicans did not.

In the wake of the 2010 midterm elections, there were at least eight major political initiatives in which members of both parties sat down to try to negotiate fiscal deals.[13] They all failed to produce results for the same reason: Republicans may have been hysterical about the alleged dangers of high deficits and a growing national debt, but they weren't prepared to work toward a compromise solution.

Helping lead the charge was Wisconsin Representative Paul Ryan, who spent the better part of a decade as a leading figure in Republican politics. But while Ryan was an aggressive deficit hawk during Obama's presidency, his trajectory on fiscal issues was emblematic of the GOP's broader incoherence on the subject.

Ryan was elected at the tail end of Bill Clinton's presidency and was generally seen as a young reactionary who grew up taking Ayn Rand novels a bit too seriously. To the extent that he was recognized at all by the political mainstream, the far-right Wisconsinite was known as a congressman who condemned Social Security as "a collectivist system" championed by Democrats, whom he dismissed as "class-warfare-breathing demagogues."[14]

When the Bush-Cheney administration advocated for a Social Security privatization scheme, Ryan not only became one of its biggest advocates, but also he wrote legislation to advance the goal. The price tag for the congressman's plan: more than $2 trillion, every penny of which would have been borrowed.[15] The Republican White House may have favored privatizing the Social

Security system, but even Bush aides concluded that Ryan's approach was "irresponsible."[16]

Under a Republican presidency, Ryan simply didn't care about the deficit. Like most of his GOP brethren, the Wisconsin lawmaker gladly went along with his party's deficit binge. In fact, he fought harder than most during Bush's presidency for his party to *deliberately* rack up far larger deficits, mocking "the green-eyeshade, austerity wing of the party," members of which "fear increases in the debt."[17] Ryan saw these GOP officials and their commitment to traditional fiscal responsibility as adhering to an antiquated and misguided ideology.

And then Barack Obama became president, at which point Paul Ryan traded his entire fiscal vision for a new one, which just happened to contradict the agenda he'd spent years supporting vociferously. The congressman's support for fiscal stimulus and deficit spending was replaced by an alternative approach that suggested fiscal stimulus and deficit spending were plagues upon society, advocated only by monsters who were trying to destroy the country.

To the consternation of those who remembered the GOP lawmaker's earlier record, the political establishment and much of the media swooned over Ryan 2.0 and his 2011 budget blueprint—known as "the roadmap"—that slashed domestic investments and privatized Medicare out of existence, ostensibly in order to eliminate the deficit.

But true to form for his post-policy party, Ryan's balanced budget plan failed to actually balance the budget. The blueprint not only relied heavily on magical assumptions[18] and wink-and-nod promises—vowing to close loopholes without identifying them,[19] for example—but also it included massive tax breaks for the wealthy, which naturally counteracted the intended savings of his proposed spending cuts.

Writing for the *New Republic* in April 2011, Jonathan Cohn crunched the numbers and found that Ryan's plan featured $4.5

trillion in reduced spending and $4.2 trillion in tax breaks.[20] The roadmap was many things, but it wouldn't guide the country to a balance budgeted.

Dismissing Ryan as "a flim-flam man," the *New York Times*'s Paul Krugman described the plan as "a fraud that makes no useful contribution to the debate over America's fiscal future."[21]

Under the circumstances, prioritizing deficit reduction in 2011 was itself a difficult goal to take seriously: unemployment was high, and previous iterations of Paul Ryan would have demanded larger deficits to spur growth. But at a more basic level, the Republican's plan was a failure on its own terms, pursuing an unnecessary goal—and poorly, too. It became painfully obvious that the man who was supposed to be the Republican Party's most insightful and knowledgeable voice of reason on budgetary issues wasn't reasonable at all.

The GOP's troubles on fiscal issues weren't limited to Capitol Hill. In August 2011, eight Republican presidential candidates stood alongside one another on a debate stage as Fox News's political anchor Bret Baier presented them with a hypothetical scenario in which Democrats offered them $10 in spending cuts for every $1 in tax increases. The moderator asked, "Can you raise your hand if you feel so strongly about not raising taxes, you'd walk away on the ten-to-one deal?"

All eight candidates raised their hands, rejecting the terms of a hypothetical deal tilted heavily in their party's favor. The commentary at the time emphasized that the presidential contenders were simply pandering to the far-right, antitax audience that had been conditioned to reflexively oppose compromises, and while there was some truth to that, it didn't negate the underlying point: these would-be chief executives were more committed to partisan posturing than to truly achieving fiscal responsibility. Anyone watching the debate wondering why bipartisan compromises on fiscal issues were impossible need only look at the eight raised hands.

The August 2011 debate was held a month after House Republicans took up a balanced budget amendment to the U.S. Constitution, which would largely prohibit federal lawmakers from spending more than the government takes in. It was curious to see some of the same members of Congress who'd added trillions of dollars to the national debt during the Bush years now take steps, under a Democratic president, to prevent Congress from running annual deficits. Just as importantly, the timing added insult to injury: in the wake of the worst economic collapse since the Great Depression, GOP lawmakers were trying to ban deficits at the exact time that larger deficit deficits made sense, in order to stimulate the moribund economy.

Deficits have routinely been used to rescue the country from dire conditions, including Franklin Roosevelt's New Deal of the 1930s and the financing of World War II. More recently, if deficits were legally impermissible, Barack Obama wouldn't have been able to end the Great Recession. A balanced budget amendment would make responses to future crises effectively impossible.

Republicans have long tried to make the case that American families have to balance their checkbooks, so the federal government should do the same. This not only reflects confusion over the dramatic macroeconomic differences between an individual family and a global economic superpower, but also it is demonstrably *wrong*: families take on worthwhile debts all the time. Americans borrow to buy a home, they borrow to put kids through college, and they borrow to start new businesses. There's nothing inherently wrong with any of this: people routinely invest in their futures, even if that means taking on debts, and it's often wise for their government to do the same thing.

It's why Bruce Bartlett, a veteran of the Reagan and Bush administrations, described the balanced budget amendment as a "dreadful" idea that was "frankly, nuts."[22]

But at its core, there was a related problem lurking just below the surface: prohibiting deficits through constitutional mandates is a lazy cop-out. If policy makers want to balance the budget, they could simply work toward that goal, as they did during Bill Clinton's presidency—a time in which the deficit was completely eliminated and operators of national debt clocks had to unplug them because the machines had never been programmed to run backward.

Republican proponents of this constitutional amendment didn't want to do the hard work of drafting responsible budgets; they wanted a gimmick to impose outcomes they otherwise wouldn't know how to reach.

Republicans didn't just look past the substantive and contextual details of fiscal debates; at times, they didn't even seem to keep up with current events. In March 2013 a veteran GOP lawmaker hosted a background briefing with journalists who could report on the discussion, so long as the member wasn't specifically identified.[23] When the conversation turned to deficit reduction, a reporter asked if it'd make a difference to congressional Republicans if Barack Obama endorsed a proposal called the chain-weighted consumer price index, or chained CPI—a change that would adjust how inflation is measured and would have the effect of reducing the growth of Social Security benefits.

"Absolutely," the lawmaker said. "That's serious."

At that point, a reporter reminded the unnamed Republican that the Democratic White House had *already* endorsed chained CPI as part of a balanced debt reduction agreement that Obama was willing to reach in exchange for comparable concessions from GOP officials.

"Who wants to do it?" the legislator asked, apparently confused. The reporter repeated that it was the Democratic president who'd grudgingly backed the idea. The longtime lawmaker, incredulous, laughed and said, "I'd love to see it."

The truth wasn't hard to find: the president's position was

available in writing on the White House's website. But the problem wasn't limited to a lazy and unobservant lawmaker: after more than two years of back-and-forth between Obama and GOP lawmakers on fiscal matters, some prominent Republican members didn't even recognize the basics of what the president had already put on the table as part of a possible deficit reduction package.

It's one thing for a party to turn down generous offers that have been proposed in writing; it's something else when a post-policy party doesn't even bother to learn what the offers are.

A month later, Congressman Greg Walden of Oregon, in his capacity as chairman of the National Republican Congressional Committee, launched an attack against the White House's stance, accusing Obama of waging "a shocking attack on seniors," and "trying to balance this budget on the backs of seniors."[24] The basis for the condemnations was simple: the president had endorsed the same chained-CPI policy Republicans had demanded he accept as part of a compromise, and now the party intended to punish him for it.

Or put another way, Republicans were going after Obama for agreeing with Republicans, accusing the president of not being liberal enough for the GOP's liking. It was an example of post-policy politics at its most farcical. Walden had effectively declared, "How *dare* the president agree with our idea that we urged him to endorse!"

Following the 2014 midterm cycle, in which Senate Republicans took the majority and House Republicans padded their existing majority, GOP officials' fiscal posture shifted further. Following six years in which Democrats made sure nearly all of their proposals were fully paid for—an era in which Republicans characterized deficit reduction as the single most important task for federal policy makers—the mask started to slip.

In the early months of 2015, the Republican-led Congress took up a series of measures—ranging from immigration, to reproductive rights, to health care—that had one important thing in common: each of the proposals would have blown up the deficit by hundreds of billions of dollars if implemented.[25]

Indeed, one of the first bills raised by the new GOP majority was a proposal to scrap the estate tax in its entirety, delivering another tax windful to the wealthiest of the wealthy. The nonpartisan Congressional Budget Office told lawmakers the legislation would add nearly $270 billion to deficits in the years that followed. Around the same time, Republicans took up the latest in a series of measures to repeal the Affordable Care Act, a move the CBO said would add more than $350 billion to annual deficits over the next decade.[26]

The party that said it prioritized about deficit reduction above all else wasted little time ignoring the CBO's findings.

Writing for the *New Republic,* Danny Vinik argued, "Republicans shouldn't be allowed to get away with this two-faced policy making. If they care about the deficit, they have to care about it in all contexts. If not, then they shouldn't justify their opposition to Obama's policies on grounds that they increase the deficit."[27] The point was more than fair, though GOP officials couldn't have cared less about what they should or shouldn't be allowed to get away with. Governing parties may concern themselves with consistent application of principles; the contemporary Republican Party would not.

One of the great ironies of the Obama era is that those who complained the loudest about the alleged scourge of budget deficits, including leaders of the so-called Tea Party movement, got much of what they wanted. Between 2009, Obama's first year in the White House, and 2015, the annual budget deficit was reduced by nearly $1 trillion. For all the far-right cries about the Democrat's approach to fiscal policy, conservatives had every reason to be impressed with Obama's results—which he achieved not with constitutional amendments, accounting gimmicks, or draconian cuts to social insurance programs, but through responsible budgeting and effective economic policies.

Obama's tenure also maintained a notable historical trend: over the last half-century, every Democratic president has left behind a deficit smaller than when he started, while every Republican chief executive has left his successor a deficit larger than when he arrived.

Donald Trump's pre-election rhetoric suggested he was prepared to interrupt the streak. In April 2016, when it was clear that the New York Republican was well on his way to winning his party's presidential nomination, Trump told the *Washington Post*'s Bob Woodward that the United States had to "get rid of" its $19 trillion national debt.[28] Asked how long he thought that would take, the future president said he could achieve the goal "fairly quickly," eliminating the debt "over a period of eight years."

It was an early indication that Trump had many things to say about the nation's fiscal challenges but had no understanding whatsoever of the relevant details. It wasn't even clear if the GOP candidate appreciated the differences between the deficit and the debt.

The words are sometimes incorrectly used interchangeably, but the deficit is an annual figure reflecting the difference when the federal government spends more than it takes in. When new deficits are added to old ones, they cumulatively become the national debt.

Throughout the Obama era, Republicans who helped create massive deficits aggressively pushed the White House to balance the budget, demanding the deficit be reduced zero. In 2016, however, Trump went much further, saying he wanted to eliminate the annual deficit *and* the $19 trillion debt, made up of previous deficits. All of this, the GOP candidate said on the record, would be possible in just two terms.

Or, put another way, Trump effectively promised to deliver an average $2.4 trillion surplus every year for eight consecutive years.

To put that in perspective, the largest surplus in modern U.S. history came in 2000, the last year of the Clinton administration, when the government took in roughly $300 billion more than it spent. Trump was suggesting he could create surpluses eight times larger than that, every year of his presidency.

The Republican also assured the electorate he could generate these results while cutting taxes, increasing military spending, and leaving Social Security and Medicare alone. As long as there have been politicians, there have been outlandish promises, but Trump's perspective on budget issues reflected a staggering display of cluelessness.

After the 2016 election, it was only natural for observers to wonder whether the Republican Party, after eight years of screaming bloody murder about the nation's budget shortfalls, would once again overhaul its entire fiscal perspective to match the partisan circumstances. During the Clinton presidency, the GOP pretended to care deeply about deficits. When George W. Bush was in office, it became "standard practice not to pay for things," in the words of Orrin Hatch. During the Obama presidency, Republicans said nothing mattered more than balancing the budget. What would the Trump era hold?

The answer started coming into focus two weeks before Inauguration Day, when members of the right-wing House Freedom Caucus, which had been adamant about reducing the deficit, announced plans to support a budget blueprint that would have dramatically increased the deficit. The *Washington Post* reported that fiscal concerns were met with a "collective shrug" from many of the conservative lawmakers.[29]

It was a sign of trouble to come. At the start of the Trump presidency, White House Budget Director Mick Mulvaney, a former fringe congressman, said he began the process of drafting a presidential budget blueprint by taking Donald Trump's speeches and "turning them into numbers."[30]

This was hilariously unwise, in part because the president's speeches tended to be incoherent, and in part because no good could come from shaping budget policy by reverse engineering stream-of-consciousness rhetoric from someone who knew practically nothing about federal budget policy.

Failure was predictable and unavoidable. Within a month of Trump taking office, the career staff at the White House Council of Economic Advisers, which is responsible for conducting academic research on economic policy for the president and his team, leaked word that Trump's budget was using ridiculously rosy estimates to obscure the fact that it would make the deficit vastly larger.[31] Indeed, the process was off to a decidedly post-policy start: instead of relying on independent data, Trump's political staff told the Council of Economic Advisers what the growth rates should be, along with instructions to use the dubious numbers to justify what Trump wanted to do.

It wasn't long before budget experts discovered that the Republican White House's blueprint applied the same $2 trillion as both the cost and the revenue of a massive package of tax cuts. A Politico analysis described it as an accounting discrepancy reminiscent of "Enron-style accounting fraud."[32] Harvard University economist Lawrence Summers, a former U.S. Treasury secretary, called it "the most egregious accounting error in a presidential budget in the nearly forty years I have been tracking them."[33] Summers added that the mistake is "a logical error of the kind that would justify failing a student in an introductory economics course."[34]

Mulvaney told reporters, "We stand by the numbers," which was in line with what one might expect from a budget director who cared as much about governing as he did about getting caught passing along an embarrassing mistake.

Cringeworthy errors notwithstanding, Americans had just experienced an eight-year period in which GOP lawmakers on Capitol Hill said they wouldn't take seriously any White House budget plan that didn't balance within ten years. Trump's didn't even try to reach this goal, and yet congressional Republicans decided collectively that it did not and would not matter.

Orrin Hatch, who was so cavalier about deficits during the Bush-Cheney years, only to change his perspective in the Obama era, reversed course once more, saying in April 2017 that he was comfortable with the Trump agenda's impact on annual deficits. "I can live with it," the Utah senator said, describing shortfalls he wouldn't stand for when a Democrat was in the White House.[35]

Hatch had plenty of company. Former New Hampshire Republican senator Judd Gregg's Obama-era panic over the nation's fiscal outlook was extreme, even by GOP standards, warning Americans of national "bankruptcy"[36] and "a fiscal crisis of potentially apocalyptic proportions."[37] In 2012 the New Hampshire Republican became the cochair of a group called Fix the Debt so that Gregg could devote more of his professional life to lobbying on behalf of a balanced budget. In August 2017, however, the former senator said it "wouldn't be so terrible" if GOP tax cuts added another couple trillion dollars to the debt.[38]

Much of the commentary of the time chided Republicans for their 180-degree turn, but the admonishments were incomplete. Clearly, the party's Trump-era posture bore no resemblance to its Obama-era positions, but this wasn't an example of a partywide flip-flop, per se. The truth was simpler: Republicans' stated concerns about deficits during Democratic administrations were little more than a fraud.

Eight years of rhetoric about a "debt crisis" came to an abrupt halt on Inauguration Day, January 20, 2017—not because the fiscal picture changed, but because the need to maintain the scam ended. To believe that the GOP changed its mind between the Obama and Trump presidencies is to believe that Republicans were occasionally sincere about their fiscal philosophy. A generation's worth of partisan posturing paints a more honest picture: those who thought the GOP seriously cared about deficits were played for fools.

For his part, Trump wasn't able to contribute much to the conversation about his party's shamelessness because he knew so little about the policies unfolding around him. In October 2017 the president suggested to *Forbes* that foreign aid was a driving force in the nation's budget shortfall. "For me, it's America first," the Republican said, peddling a mantra with no practical meaning. "We've been doing that so long that we owe twenty trillion, okay?"[39]

What the president didn't understand is that the United States spends less than $50 billion a year on foreign aid. For the typical person, that's an unfathomable amount of money, but in the context of the U.S. budget, it's practically a rounding error. To see this part of the budget as a major contributor to the national debt is difficult to take seriously.[40]

Around the same time, Trump told Fox News that he was reducing the national debt "in a sense" as the stock market rose. "We picked up $5.2 trillion just in the stock market," Trump told Sean Hannity, pointing to a number that was itself dubious. "So you could say, in one sense, we're really increasing values. And maybe in a sense, we're reducing debt. But we're very honored by it."

These were the comments of a man who didn't understand how money works. Unless the White House intended to seize all of the profits of every Wall Street investor, the national debt doesn't shrink when Wall Street indexes rise or fall. Trump wasn't reducing debt "in a sense"; he was adding trillions to the debt in reality.

As 2017 came to a close, independent budget scorekeepers told GOP policy makers the party's regressive tax plan, approved without hearings or scrutiny, would add at least $1 trillion to future deficits. Republicans ignored the warnings, confident in the knowledge that tax breaks for the wealthy pay for themselves through increased economic dynamism—an absurd assertion generally known as believing in the "tax fairy."

The more Republicans shift into post-policy politics, the less interest they show in evidence, but the idea that tax cuts can pay for themselves has been the subject of extensive study. The results don't change: the idea doesn't work. The *New York Times*'s Paul Krugman explained in June 2019: "No idea in economics has been as thoroughly tested—and as completely rejected—as the notion that tax cuts pay for themselves. The reason it has been tested so much is that Republicans keep insisting that it's true, and base policy on the claim. Then, when it fails—which it has done again and again, from Bush to Trump, from DC to Kansas—they pretend not to notice and do it all over again."[41]

The *Times*'s Jim Tankersley added, "It's time to put to rest any notion that President Trump's signature tax cuts are paying for themselves. Anyone who says otherwise is lying with numbers."[42]

A governing party would recognize the evidence and adjust accordingly. Republicans didn't like the evidence, so they chose not to believe it.

In the political equivalent of thumbing their nose at responsible governing, House Republicans tried once again to pass the balanced budget amendment to the Constitution in April 2018—just four months after the same GOP lawmakers passed unnecessary tax breaks for the wealthy, in the process pushing the deficit to a seven-year high.

Three months later, Larry Kudlow, the director of the White House's National Economic Council, told the Fox Business Network, "As the economy gears up, more people working, better jobs and careers, those revenues come rolling in, and the deficit, which was one of the other criticisms, is coming down. And it's coming down rapidly."

It was up-is-down politics at its most exasperating: at the time of Kudlow's comments, the deficit that his boss had vowed to eliminate was going up rapidly. The top voice on economic policy in the White House was either embarrassingly confused or actively engaged in an ugly campaign to gaslight the public.

Either way, the comments reflected the perspective of someone who plainly didn't care about the relevant facts and wasn't prepared for a real debate over policy.

By the time Trump was preparing to deliver his State of the Union address in early 2019—a speech that ignored fiscal issues altogether—the circle was complete: Mick Mulvaney, an unrestrained, hair-on-fire deficit hawk during Obama's presidency, had been elevated to serve as acting White House chief of staff and was letting his GOP allies know that "nobody cares" about the deficit anymore.[43]

The cynicism was breathtaking. Mulvaney used to boast that he got involved in politics in part because he disapproved of the Bush-Cheney administration's big budget deficits.[44] In 2011 the far-right South Carolinian even helped lead the Republican charge on the party's debt ceiling crisis, arguing, "Anybody who is up to speed on budget issues should be scared to death by what's happening with the debt and the deficit in this country. If you're not losing sleep over it, then you're simply not paying attention."[45]

Once a GOP president was in the White House, and the national debt passed the $22 trillion mark, Mulvaney acted as if "paying attention" was no longer worthwhile.

Similarly, Larry Kudlow said in July 2019, as the national debt reached $22.5 trillion, fueled by tax cuts Republicans couldn't pay for, "I don't see this as a huge problem right now at all."[46] A decade earlier, however, when the debt was roughly half the size, Kudlow was on the verge of panic about "humongous deficits."[47]

For the White House's critics, it created a target-rich environment. There was the hypocrisy angle: Republicans spent the Trump era running up massive deficits, doing the opposite of what they'd said they believed throughout Obama's presidency. And then there was the dishonesty angle: Trump said he could eliminate the deficit and pay off the debt, and he wasn't even trying to honor his commitments to voters.

But the post-policy angle was just as overt: Republican officials in the White House and on Capitol Hill acted as if they didn't know or care about arithmetic, economic policy, or the real-world implications of their halfhearted attempts at governing. Told that their tax plan would immediately cause the deficit to soar, GOP lawmakers shrugged. Told later that their tax plan had, in fact, caused the deficit to soar, Republicans didn't bother to admit they were wrong or take even the most modest steps to undo what they'd done.

Reminded that their actions were wholly at odds with their own purported philosophy, GOP officials could barely be bothered to lift an eyebrow.

The idea of making serious attempts at governing on the issue simply never occurred to Republicans. By all appearances, they wouldn't have even known how to try.

CHAPTER 8

Life and Death in the Culture Wars

Gun Control, Civil Rights, Reproductive Rights

IN THE WAKE OF a mass shooting at Umpqua Community College in Roseburg, Oregon, in 2015, which left ten dead and eight injured, Barack Obama responded with a familiar degree of frustration, though he also touched on an underappreciated angle to the debate over gun violence.

"We spent over a trillion dollars and passed countless laws and devote entire agencies to preventing terrorist attacks on our soil, and rightfully so," he said. "And yet we have a Congress that explicitly blocks us from even collecting data on how we could potentially reduce gun deaths. How can that be?"

It wasn't a rhetorical question, and there was nothing mysterious about the answer. In the 1990s, the Clinton administration's Centers for Disease Control and Prevention expanded its research into gun-related deaths as a public health issue, which didn't sit well with congressional Republicans. Representative Jay Dickey of Arkansas, with the support and assistance of gun industry lobbyists, added language to the appropriations bill that financed the

CDC: "None of the funds made available for injury prevention and control at the Centers for Disease Control and Prevention may be used to advocate or promote gun control."[1]

The Dickey Amendment, as it came to be known, was so vague, scholars had no idea what kind of research might be construed as advocating or promoting firearm restrictions, and the result was a broad chilling effect: the CDC and its partners steered clear of the issue altogether, fearing what might happen if GOP lawmakers decided their scientific pursuits crossed a nebulous legal line.

Scholars in the field felt constrained, even as deadly mass shootings continued. The *Los Angeles Times* reported in September 2019 on every public mass shooting since 1966 for a project funded by the National Institute of Justice, which serves as the research arm of the U.S. Department of Justice, and found that the attacks "are becoming far more frequent, and they are getting deadlier."[2]

The reporting added, "During the 1970s, mass shootings claimed an average of 5.7 lives per year. In the 1980s, the average rose to 14. In the 1990s, it reached 21; in the 2000s, 23.5. This decade has seen a far sharper rise. Today the average is 51 deaths per year." The statistics lent credence to the plain fact that the political debate over gun policy and gun violence is far more than a culture war issue. It's also a public health issue.

Nevertheless, for much of the last generation, the body of empirical scientific evidence on gun violence dried up—by Republicans' design. GOP officials didn't want there to be evidence-driven policy making on the issue, so they took steps to ensure there was no evidence upon which to govern. It wasn't until December 2019, nearly a quarter century after the Dickey Amendment was first approved, that Congress agreed—at the House Democratic majority's insistence—to fund research into gun violence.[3]

It's hardly a secret that the contemporary Republican Party, closely aligned with right-wing advocacy groups such as the National Rifle Association (NRA), is a fierce opponent of every proposed restriction on firearms. GOP officials, fueled by a bizarre and paranoid fantasy that an unarmed population would be subjugated by tyrants, have invested considerable energies into arguing that the only thing standing between contemporary American society and a totalitarian state is policy makers' fears of an armed uprising.

What's less appreciated is the degree to which GOP officials approach the culture war debate in a decidedly post-policy way, making a substantive debate all but impossible and leaving sweeping societal consequences in their wake.

In early 2013, for example, as the nation was still coming to grips with the massacre at Sandy Hook Elementary School in Newtown, Connecticut, Senate Minority Leader Mitch McConnell's reelection campaign sent a written message to supporters condemning presidential executive orders that didn't exist and rejecting the idea of universal background checks as a "thinly veiled national gun registration scheme" intended to "ensure federal government minders gain every bureaucratic tool they need for full-scale confiscation."[4]

McConnell's team couldn't have believed such claims, but he and his allies knew that a meaningful debate based on the facts would likely lead to reforms that many Republican policymakers and their allies wouldn't like. It therefore became necessary to shape the policy fight in such a way as to ensure that the substance was irrelevant.

In any major legislative dispute, political players invariably try to persuade the public to agree with them. In the spring of 2013, as reform advocates rallied behind proposals for universal background checks on gun purchases, this aspect of the overall initiative proved to be the easiest: a *Washington Post*–ABC News poll asked respondents, "Would you support or oppose a law requiring background checks on people buying guns at gun shows?" To understate matters, the results were one-sided: 91 percent of Americans supported the idea.[5]

Given widespread public divisions on a whole range of political issues, the data suggested that mandatory background checks were among the single most popular proposals in the United States. There was, in effect, a public consensus.

"How often do ninety percent of Americans agree on anything?" Obama said to laughter at an event in Connecticut in April 2013. "And yet, there is only one thing that can stand in the way of change that just about everybody agrees on, and that's politics in Washington. You would think that with those numbers, Congress would rush to make this happen. That's what you would think. If our democracy is working the way it's supposed to, and ninety percent of the American people agree on something, in the wake of a tragedy, you'd think this would not be a heavy lift."

But it was a prohibitively heavy lift, thanks in large part to congressional Republicans arguing that a system of universal background checks would lead to a database that would ultimately serve as a federal registry of gun owners. The architects of the Senate background-check bill, eager to address those concerns, explicitly prohibited the existence of such a database.[6] In fact, the text of the legislation included a provision that read, "Nothing in this title, or any amendment made by this title, shall be construed to . . . allow the establishment, directly or indirectly, of a Federal firearms registry." Under the same bill, anyone caught trying to create a federal registry would be charged with a felony, subject to fifteen years behind bars.

Republicans couldn't come right out and admit what was plainly true—they were going to oppose all reforms, in part because Barack Obama supported them, and in part at the insistence of the NRA—so they created an alternate reality in which a popular bill included radical provisions that didn't exist. There was nothing proponents of the legislation could do to address GOP lawmakers' objections because their concerns were inherently insincere. Reform advocates came to realize there was no way to satisfy the demands of those arguing in bad faith because they didn't want to be satisfied.

Almost exactly four months after the Newtown massacre, in which a lone gunman wielding a semiautomatic AR-15 rifle shot to death twenty young children and six adults inside a school before turning the gun on himself, the background-check bill failed. It enjoyed majority support in the Democratic-led Senate, but the proposal needed sixty votes to overcome a Republican filibuster, and it ended up with fifty-four.

A week later, Republican senator Pat Toomey of Pennsylvania, who'd helped coauthor the background-check bill, conceded, in reference to his party, "There were some on my side who did not want to be seen helping the president do something he wanted to get done, just because the president wanted to do it."[7]

In the wake of the bill's defeat, there was some media chatter that the president bore responsibility because he hadn't twisted enough arms or "led" in sufficiently effective ways. But Toomey's quote helped underscore the reality of the unyielding circumstances: Republicans had no interest in governing. No amount of cajoling could overcome the will of lawmakers who wanted to defeat the president's agenda because it was the president's agenda.

As the 2016 presidential race got under way, the debate descended further. Senator Marco Rubio of Florida, who helped lead the fight against background checks, argued during a September 2015 Republican debate, "The only people that follow the law are law-abiding people. Criminals by definition ignore the law, so you can pass all the gun laws in the world, like the left wants. The criminals are going to ignore it because they are criminals."[8]

Of course, by that reasoning, there's no real point in having any law addressing any societal ill. The only people who follow laws against stealing cars, for example, are law-abiding people. Liberals can pass all the laws they want against car thefts, but if the Republican senator were right, those laws serve no real purpose, since car thieves will ignore those laws because they're criminals.

A few months later, Rubio made the case on Fox News that gun-safety measures wouldn't save lives, and he was annoyed by the left's overly narrow focus: "I don't hear anybody talking about bomb control."[9] What the conservative Floridian seemed wholly unaware of was the fact that the United States already had extensive laws limiting public ownership of bombs, which helped explain why domestic mass murders using explosive devices are quite rare.

Daily Kos, a leading progressive blog and online community, published an item in response to Rubio, patiently trying to bring him up to speed on the public policy he claimed to care about but hadn't bothered to research: "You are not allowed to own a bomb, or carry a bomb on your back while visiting your neighborhood Chipotle. Not even if you claim you need a bomb for self-protection, or because you're afraid there are other people out there with other bombs who might try to bomb you first. . . . Even those Americans who are permitted to own and operate bombs (for example, for excavation purposes) face very, very tight state and federal regulations and restrictions; as for the rest of us, outside of military service we are allowed to be in personal possession of a mass-murder-sized bomb approximately never."[10]

Rubio wasn't the only GOP presidential hopeful who didn't know what he was talking about. In October 2015 Donald Trump headlined an event in South Carolina where he described a threat from the Obama White House he'd created out of whole cloth. "You know, the president is thinking about signing an executive order where he wants to take your guns away," the Republican told voters. "You hear about this?"[11]

No one had because there was no such plan, and there is nothing in federal law that would empower any president to confiscate guns through executive action.

Once in the White House, Trump's understanding of the issue seemed to deteriorate further. In early 2017, one of the very first measures the new president signed into law was a bill to, of all things, expand gun access to the mentally impaired.

When an American suffers from a severe mental illness, to the point that he or she receives disability benefits through the Social Security Administration, there are a variety of safeguards in place to help protect that person and his or her interests. These Americans cannot, for example, go to a bank to cash a check on their own.

Before 2017, the mentally impaired weren't supposed to be able to buy a gun, either. Officials established a fairly straightforward system: the Social Security Administration would report to the FBI's background-check system the names of those who receive disability benefits due to severe mental illness.

At the behest of congressional Republicans, Trump signed into law legislation to sever the connection, blocking the Social Security Administration's reporting policy. As for why he supported the change in policy and made it one of the very first policies he tackled after taking office, no one could say for sure. Trump said effectively nothing about it, and there was little to suggest he even knew what he was signing.

Nevertheless, the president and other proponents responded to several mass shootings by stressing the importance of keeping guns from those with severe mental illnesses. Iowa senator Chuck Grassley said, for example, "We have not done a very good job of making sure that people that have mental reasons for not being able to handle a gun, getting their name into the FBI files, and we need to concentrate on that."[12] Grassley was the chief sponsor of the Senate bill that kept the names of the severely mentally ill off the FBI background-check list.[13] He made no effort to explain the contradiction.

Toward the end of Trump's first year in office, a gunman in a thirty-second-floor hotel suite opened fire on an outdoor music festival audience in Las Vegas, killing fifty-eight people and wounding hundreds more in the deadliest mass shooting in American history. When Senator James Inhofe was asked a day later about the prospect of legislation to help keep Americans safe from future incidents, the Oklahoman captured the vapidity of the policy debate beautifully: he said the prevalence of "sanctuary cities" led to a sense of lawlessness, which in turn had "a profound impact on people's behavior."[14]

The same week, South Dakota senator John Thune, the chairman of the Senate Republican Conference, added that Americans looking to policy makers to approve major new public-safety measures should probably lower their expectations. "I think people are going to have to take steps in their own lives to take precautions, to protect themselves," he said. Referring to the Las Vegas slayings, Thune added, "In situations like that, you know, try and stay safe. As somebody said—get small."[15]

Comments such as Thune's and Inhofe's were foolish to the point of being insulting, but they were also emblematic of a larger truth: presented with a deadly national scourge, GOP officials didn't want to try to govern or even come up with compelling excuses for their passivity.

A month after the massacre in Las Vegas, there was a mass shooting at a church in Sutherland Springs, Texas, claiming twenty-six lives. The next morning, Trump said the murders didn't constitute "a guns situation."[16] For Republicans, they never are. In response to nearly all domestic mass shootings, the party lazily rolls out a series of tired talking points, chief among them the idea that video games are responsible for bloodshed. "We must stop the glorification of violence in our society," Trump said after one of the many massacres that occurred during his tenure. "This includes the gruesome and grisly video games that are now commonplace."[17]

It was a familiar refrain. A couple of months after the massacre at Sandy Hook, Republican senator Lamar Alexander of Tennessee argued on MSNBC, "I think video games is [*sic*] a bigger problem than guns, because video games affect people."[18]

In theory, GOP policy makers could have reviewed the extensive social-science research that's publicly available, which makes clear that playing violent video games does not lead to violent crimes.[19] Republicans also could have evaluated crime rates in other countries, where people play the exact same titles and gaming is a huge cultural phenomenon, but where mass shootings are unheard of.

But those details are ignored, not because they're unreliable but because the underlying argument is itself insincere. If Republicans were seriously concerned about the societal impact of video games, they'd do more than briefly drag out this talking point in the immediate aftermath of deadly slayings. It's a cliché intended to distract those who expect GOP officials to act like members of a governing party.

The broader national debate took another turn in February 2018, when a gunman using a semiautomatic rifle killed seventeen people in a Parkland, Florida, high school. Within days of the bloodshed, the president argued via Twitter that the FBI might have prevented the mass murders if federal investigators had only stopped investigating the connections between Trump's political operation and its Russian benefactors during the 2016 presidential election that put him in the White House.[20]

The presidential whining was as callous as it was ridiculous, but it also helped touch off one of the more dramatic debates of Trump's presidency.

As public demand for action grew, the White House concluded that ignoring the issue altogether wasn't a credible option. A week after the massacre in Parkland, the president hosted an event on school shootings, at which Trump presented his audience with his preferred strategy: having a variety of authorized personnel with guns in schools: teachers, coaches, principles, and so on. He said the approach "could very well solve your problem."

The president arrived at this solution the same way he considered every other policy challenge: by going with an unexamined, overly simplified answer that appealed to his version of common sense. The immigration system is broken? Build a wall. Opioids are ravaging communities? Execute drug dealers. Hurricanes are approaching American soil? Hit them with nuclear weapons.[21] There are too many shooters killing children in schools? Put more guns in the hands of those who might shoot back. Solutions like these aren't bolstered by analyses or scholarship because, in a post-policy model, that's simply not how questions are answered.

At the White House, where the president touted this approach, he asked the audience if anyone liked the idea. Practically no one raised their hands. Trump then asked if anyone in the room was "strongly against" the idea, and many attendees were quick to raise their hands.

Almost immediately, the chief executive decided to run with his proposal anyway. When this sparked pushback, he defended the idea by insisting that "twenty percent of teachers" are "adept" with firearms, and would therefore be prepared to engage a gunman and neutralize him in the event of a school shooting. The Republican added on Twitter, "ATTACKS WOULD END! . . . Problem solved."[22]

The president's confidence notwithstanding, this was an idea he really should have thought through before sharing with the public. Putting aside obvious questions about keeping children safe from guns by surrounding them with more guns—something that didn't work in Parkland, where there was an armed officer who neither stopped nor deterred the violence—there were practical considerations Trump hadn't stopped to consider.

For example, in a country with 3.6 million teachers, arming 20 percent of them would require purchasing and distributing more than 700,000 firearms—a total that would grow considerably if coaches, administrators, and principals were added to the mix. This may have been a dream for the NRA, which exists largely to protect the interests of gun manufacturers, but the idea that the United States would spend $1 billion on such a policy was difficult to take seriously.[23]

Just as importantly, Trump's alleged statistic that one in five American schoolteachers were "very gun-adept people" was hopelessly bonkers. Vox published a detailed analysis showing that the actual number was vastly smaller, and quoted Diane Ravitch, an educational policy analyst at New York University and a former U.S. assistant secretary of education, who said the president simply "made up" the statistic.[24]

The White House responded to these concerns by ignoring them, because none of the president's ideas about mass shootings suggested a serious attempt at governing. It's not as if officials in the West Wing were eagerly distributing detailed fact sheets about the real-world logistics of Trump's proposal. For all intents and purposes, there was no proposal. There were tweets about an absurd idea from a president who didn't know what he was talking about and wasn't prepared to spend time getting up to speed.

In late February 2018 Trump took the discussion, such as it was, in an even stranger direction, holding a televised, hour-long discussion with a group of lawmakers from both parties and both chambers of Congress. Operating on impulse and instinct, the president mocked his ostensible Republican allies for being "afraid" and "petrified" of the NRA. "They have great power over you people," the president told members of his own party, in reference to gun industry lobbyists. "They have less power over me."

Soon after, at the same meeting, Trump voiced support for extrajudicial gun confiscations. Vice President Mike Pence raised the prospect of empowering law enforcement to take weapons away from those who've been reported to be potentially dangerous, though he added that he expected to see "due process so no one's rights are trampled." The president wasn't satisfied with the answer.

"Or, Mike, take the firearms first and then go to court," Trump interjected, adding, "To go to court would have taken a long time.

You could do exactly what you're saying, but take the guns first, go through due process second." At the same event, the president endorsed law enforcement models in which police officers confiscated some Americans' guns "whether they had the right or not."

Trump went on to endorse a "comprehensive" package of gun reforms that included background checks and raising the minimum age for gun purchases to twenty-one. "Now, this is not a popular thing to say, in terms of the NRA, but I'm saying it anyway," the Republican boasted, predicting that such a bill could clear the Senate with more than sixty votes.

A GOP staffer on Capitol Hill told CNN after the gathering, "It feels very much like nobody briefed the president before this meeting."[25] It was a polite way of saying Trump spent an hour talking about addressing gun violence in ways he didn't understand, touting an agenda neither his team nor his party had any intention of supporting.

Another Republican staffer told the conservative political magazine the *Weekly Standard*, "This is why you don't do high-stakes, hot-button negotiation on live TV with someone who doesn't know or care about details."[26]

It wasn't just a throwaway line. In the American system of government, it's supposed to matter what a sitting president believes about pending policy proposals, especially concerning life-or-death issues. In fact, in some cases, success hinges on whether a president can be persuaded to support or oppose specific ideas. But in the Trump era, the executive branch was led by someone who didn't "know or care about details," which in practical terms meant there wasn't any point in having policy discussions with the post-policy president on this or any other subject.

Officials had to adjust to a governing dynamic—or something resembling a governing dynamic—in which the sitting American president didn't necessarily know what he was saying at any given moment, would routinely contradict his own administration's positions, and with even greater frequency, would radically change direction on his positions, often in response to something he saw, and only slightly understood, on cable news. It made negotiating with the president, as New York Democrat Chuck Schumer famously put it, like "negotiating with Jell-O."[27]

Trump had plenty of time to prepare for the meeting with lawmakers on gun policy; he simply chose not to brush up on the subject at hand.

A day later, the president welcomed NRA representatives into the Oval Office. Soon after the private discussion, a top lobbyist for the right-wing group announced that Trump had retreated from the statements he'd made to lawmakers twenty-four hours earlier. White House Press Secretary Sarah Huckabee Sanders told reporters that the president only agreed "conceptually" with his stated positions.[28] The legislative efforts evaporated soon after.

Those who believed the president's rhetoric on comprehensive gun reforms had learned a valuable lesson: Trump was too often an unreliable narrator about his own plans.

As his presidency advanced, his willingness to take the issue seriously did not improve. In May 2018 Trump spoke at an NRA event and said, "Your Second Amendment rights are under siege, but they will never, ever be under siege as long as I'm your president."[29] The fact that the first half of the sentence was at odds with the second was a detail the Republican chose to ignore.

Trump added, "It seems that, if we're going to outlaw guns, like so many people want to do . . . we are going to have to outlaw, immediately, all vans and all trucks, which are now the new form of death for the maniac terrorists." What the president didn't seem to realize is that vans and trucks are already heavily regulated. Indeed, van and truck drivers are required to have licenses and insurance; they're tested by the state before they can legally hit the road; and it's standard practice for government agencies to maintain extensive ownership and registration records, just as a matter of course.

Or put another way, Trump hadn't thought through the substance of the analogy.

In early 2019 the newly elected House Democratic majority easily passed a bill to expand federal background checks on gun purchases—the first major reform of American gun laws to pass either chamber of Congress in over a decade. The White House issued a formal veto threat, vowing to reject the legislation if it ever reached the president's desk.

Eight months later, following a pair of deadly mass shootings over a single weekend in August 2019, Trump rediscovered his support for "strong" background checks.[30] According to multiple accounts, Trump was excited by "a Rose Garden fantasy" in which he'd sign sweeping gun reforms, boosting his reelection prospects, strengthening his party's weakening standing among suburban voters, and delivering on a popular priority on which his predecessor couldn't muster the congressional support.[31]

Asked during a Q&A with reporters whether his far-right supporters would go along with such a change, the president said he wasn't concerned—because members of his base "rely on me in terms of telling them what's happening."

As reform advocates felt a glimmer of hope, it was hard not to think of poor Charlie Brown, standing on a field, wondering if this would be the time Lucy finally let him kick the football without yanking it away at the last moment.

Alas, Charlie ended up on his backside once more. As had happened before, the NRA dictated the outcome: Trump had a thirty-minute phone call with Wayne LaPierre, the powerful organization's chief executive, shortly after he'd made his public statements favoring some form of gun control.[32] And, predictably, just like that, the president once again retreated from the policy he said he supported.[33] Background checks? Off the table.

The broader American culture war has long featured an unsettling number of skirmishes that have nothing to do with guns, though they're bound by a common theme: Republican disdain for the substance of governing.

The party's tactics in opposition to civil rights for the LGBTQ community shared the same characteristics. In Barack Obama's first year in office, the White House prioritized ending the military's Don't Ask, Don't Tell (DADT) policy that prevented gay Americans from serving openly in the military. John McCain, a decorated war hero who said he'd served alongside gay servicemen, positioned himself as the chief opponent of the Democratic effort.

"My opinion is shaped by the views of the leaders of the military," the senator from Arizona said in June 2009. "The reason why I supported the policy to start with is because General Colin Powell, who was then chairman of the Joint Chiefs of Staff, is the one that strongly recommended we adopt this policy in the Clinton administration. I have not heard General Powell or any of the other military leaders reverse their position."[34]

McCain wasn't paying close enough attention. Powell had backed off his support for DADT seven months earlier, and his successor, retired army general John Shalikashvili, also urged policy makers to end the discriminatory policy. It was around this same time that twenty-eight retired generals and admirals, many of whom supported the DADT policy when it was first enacted, said they believed it was time to repeal the law.[35]

As the debate progressed, McCain moved the goalpost. It wasn't enough for Colin Powell to have changed his mind; the senator also wanted to hear from the Pentagon's senior leadership. When Secretary of Defense Robert Gates also endorsed repeal, McCain said the Pentagon chief wasn't an actual military leader, so his judgment wasn't persuasive. When Joint Chiefs chairman Admiral Mike Mullen also endorsed ending the ban on openly gay troops, McCain said that wasn't good enough, either; he'd need to see the results of a Pentagon survey of active duty personnel and their families.

In November 2010 the Department of Defense released the results of its yearlong study, which found that more than 70 percent of respondents were comfortable with ending Don't Ask, Don't Tell.

McCain shifted gears again, saying *that* wasn't enough, either, and he'd need another year to scrutinize the survey's findings. The Arizona Republican, in one of the last high-profile legislative fights of his lengthy career, seemed reluctant to acknowledge what was plainly true: he simply didn't want openly gay service members in the military, and his fierce opposition wouldn't waver, regardless of the evidence.

Every question McCain and his GOP allies raised was answered. Every concern was addressed. Every scary scenario was examined and discredited. Ultimately, the repeal measure passed the Senate, though three out of four Republicans in the chamber voted against it.[36]

Societal progress on this issue advanced throughout the Obama era, objections from the right notwithstanding. Not only did Democrats end the ban on gay and lesbian soldiers serving openly, but also they helped confirm an openly gay man, Eric Fanning, to serve as the secretary of the Army. The Obama administration also ended the ban on transgender Americans serving in uniform. Defense Secretary Ash Carter announced in June 2016 that transgender service members are "talented and trained Americans who are serving their country with honor and distinction." The change in policy proceeded without incident.

That is, until July 2017, when Donald Trump published a few tweets announcing that he would no longer "accept or allow transgender individuals to serve in any capacity in the U.S. military." Within hours, the White House was framing the shift in an overtly political way, with one administration official arguing, "This forces Democrats in Rust Belt states like Ohio, Michigan, and Wisconsin, to take complete ownership of this issue."[37] It was hard not to get the impression that military recommendations, honoring American service men and women, and civil rights were not the president's principal concerns.

Just as importantly, Trump arrived at this dramatic shift in policy without having done any meaningful work. The same day as the presidential tweets, Pentagon officials said they'd been completely blindsided by Trump's announcement. Not only weren't they consulted about the ban on thousands of American troops, but also they weren't even given a heads-up that the tweets were coming. A spokesperson for the Department of Defense refused to answer reporters' questions on the new policy, leaving it to the White House to respond to inquiries military officials knew nothing about.[38]

No one could explain why Trump embraced the discriminatory new policy, how he arrived at it, or exactly what problem the president thought he was solving. In the two weeks following the tweetstorm, the Pentagon and the service chiefs were unsure how, or even whether, to proceed with discharging active duty personnel, and some military officials largely ignored Trump's announcement, pending some kind of policy review.

A reporter asked the president in August 2017 why in the world he did this. After bragging about his belief that he received a lot of votes from military personnel, Trump said, "It's been a very difficult situation. And I think I'm doing a lot of people a favor by coming out and just saying it. . . . I think I'm doing the military a great favor."

It was a favor the military had not requested and barely understood. Navy Secretary Richard V. Spencer, a Trump nominee, took issue with his boss's policy, arguing, "On a fundamental basis, any patriot that wants to serve and meets all the requirements should be able to serve in our military." Coast Guard Commandant Admiral Paul Zukunft pledged he would not "break faith" with the transgender service members already in uniform.

Joseph Dunford, the chairman of the Joint Chiefs of Staff, left the status quo in place for months while the White House scrambled to formulate an actual policy to follow—which reinforced impressions that the military didn't see the president's directive as "a great favor."

More than a year later, after the U.S. Supreme Court's conservative justices allowed the discriminatory policy to be implemented, Trump still didn't fully understand what he'd done or why. In June 2019, the president sat down for an interview with Piers Morgan, a British broadcaster, who asked Trump to explain his discriminatory approach.

"Because they take massive amounts of drugs, they have to—and also, and you're not allowed to take drugs," Trump said, in apparent reference to transgender members of the military. "And they have to after the operation. They have to. They have no choice. And you would actually have to break rules and regulations in order to have that."

When the host reminded the Republican that the cost of the medications is miniscule, and that the Pentagon already spent far more to provide servicemen with the erectile dysfunction drug Viagra, Trump's defense of his policy came down to this: "Well, it is what it is."

His blithe indifference to the details was emblematic of his broader ignorance. The administration had argued in court that the ban on transgender Americans serving in the military was needed to protect unit cohesion and morale. Trump discarded this talking point to insist his policy was related in some incoherent way to medications.

Except this was factually wrong, too. Military personnel routinely take prescription medications, and the Pentagon makes the treatments available to troops around the world every day. That includes hormone therapies—some of which have nothing to do with gender identity treatments—which are already covered by existing Pentagon guidelines.

Indeed, the Defense Department, which is not generally in the habit of publicly contradicting the commander in chief, quickly went on the record to assure the public that the president was wrong. "The Military Health System covers all approved medically necessary treatments and prescription medications," DOD spokeswoman Jessica Maxwell told reporters in a written statement. "If a service member has a hormone deficiency for any reason (such as hypogonadism, hypothyroidism, menopause, etc.), he or she would be prescribed hormones."

What's more, as a *Washington Post* analysis explained, "It's also important to note that not all transgender people undergo gender reassignment surgery or take prescription hormones, so even if such prescribed drugs were prohibited, it wouldn't necessarily mean transgender troops would have to be banned from serving. Trump's comments seem to suggest that these drugs would be required for all transgender troops; they're not."[39]

The president's comments came just under two years after his original tweets announcing the ban, which meant he had nearly two full years to come up with an argument to defend his discriminatory policy. For all of Trump's bluster about his love and respect for the military, and despite his campaign assurances that he would stand up for the LGBTQ community, in this case, he didn't care enough to bother.

Few issues motivate Republicans like abortion and reproductive rights. In much of the country, it's a litmus test issue.

As the 116th Congress got under way in 2019, the number of pro-choice lawmakers in the House GOP was down to literally zero—and the party's positions on abortion and reproductive health have gradually become a pillar of Republicans' national identity, driving the party on everything from judicial nominees, to candidate selection, to electoral fund-raising.

If the GOP succeeds on this front, millions of women would

not just lose the ability to legally terminate unwanted pregnancies, but also the impact would be far broader and encompass a wide range of issues.

It makes it all the more important, therefore, to appreciate the degree to which Republicans approach these fights in a decidedly post-policy way.

For example, in 2011 Texas governor Rick Perry sat down for an interview with *Texas Tribune* editor Evan Smith, who passed along a question from the public. A voter wanted to know, "Why does Texas continue with abstinence education programs, when they don't seem to be working?" The question was well grounded in fact: in the areas of teen pregnancies and teen births, Texas ranked among the worst states in the United States.

Perry heard the question, thought for a second, and replied, "Abstinence works."[40]

The incredulous reporter pressed on, reminding the governor, "But we have the third-highest teen pregnancy rate among all states in the country. The questioner's point is, it doesn't seem to be working." The governor responded again, "It, it works."

The exchange was a post-policy classic: the Texas Republican's policy had, in quantifiable ways, been discredited as a failure, at both the state and national level. Presented with the results, Perry nevertheless insisted that the policy was effective, reality be damned, because in his approach to governance, believing in a bad idea was all the proof he needed of its efficacy.

Perry's GOP brethren weren't much better. The same year, as GOP lawmakers waged a bitter fight to defund Planned Parenthood, Senate Minority Whip Jon Kyl of Arizona defended his party's tactics by claiming that abortion services made up "well over 90 percent of what Planned Parenthood does." That wasn't even close to being true: abortion services represent a small fraction of the health care organization's overall work.

When reporters called the senator's office seeking clarification,

Kyl's spokesperson said the Republican's remark "was not intended to be a factual statement."[41] It was unclear how Democratic and GOP policy makers were supposed to work cooperatively on this or any other issue when one side uses made-up statistics that aren't intended to be factual.

The Republican crusade against Planned Parenthood took a more hilarious turn in 2015 when the GOP-led House Committee on Oversight and Reform, chaired by Utah GOP representative Jason Chaffetz, held a hearing and presented Planned Parenthood's then-president, Cecile Richards, with a bizarre chart purporting to show a sharp increase in the number of abortions the organization had performed, coinciding with a sharp decrease in cancer screenings and preventive services.[42]

All things considered, the image was more of an inept art project than a graph intended to convey meaningful information: it had no axes, and the data points seemed to have been inserted at random. Chaffetz claimed he pulled the figures from Planned Parenthood annual reports. That wasn't true: as Richards reminded the congressman, the image the Utah Republican was pointing to literally said, in all capital letters, "Source: Americans United for Life."

In other words, Chaffetz relied on a far-right antiabortion organization's bogus chart, pretended the careless, axis-free image was his own creation, and proceeded to humiliate himself at a hearing he called but failed to prepare for. When the Planned Parenthood president turned the tables on the confused committee chairman, he sat silently for a moment, displayed a deer-in-the-headlights look, and scrambled to move on.

If Republicans on the committee had taken the substance of the debate more seriously, the embarrassing stunt, which backfired in predictable ways, could have been avoided.

While GOP officials went after Planned Parenthood at the institutional level, they also took on a related campaign in support of medically unnecessary ultrasound procedures. The unprecedented effort was predicated on the idea that if a woman seeking to terminate an unwanted pregnancy was forced by law to undergo an ultrasound scan, she might change her mind about having an abortion after having seen the image of the fetus.

During the debate over the Affordable Care Act, Republicans insisted that any political interference with the doctor-patient relationship was intolerable. The party nevertheless championed a series of state-based laws that required women to undergo procedures they did not want or need and their physicians saw no reason to provide.

In Pennsylvania, Governor Tom Corbett, a Republican, said that women who objected to state-imposed medically unnecessary ultrasounds should "close their eyes" during the procedure.[43] Kentucky governor Matt Bevin, also a Republican, went further: doctors who perform abortions in the state were legally obligated to perform ultrasound procedures and describe the on-screen images to women, in detail, before terminating an unwanted pregnancy.[44] Patients who objected were required to shield their eyes and make self-generated noises to drown out the information physicians were forced by Kentucky politicians to share against their will.

A variety of phrases come to mind to describe efforts like these. "Responsible policy making from a governing party" isn't among them.

Wisconsin governor Scott Walker seemed especially proud of his law mandating medically unnecessary ultrasounds, which required medical professionals to describe the on-screen images in detail, whether they wanted to or not. The governor told reporters in 2013, "I don't have any problem with ultrasound."[45]

In reality, the question wasn't whether ultrasounds are a legitimate diagnostic tool that can play a meaningful role in modern medicine; the question was whether GOP politicians should be mandating that patients undergo ultrasound procedures for no medical reason whatsoever, regardless of the patient's wishes, regardless of what medical professionals considered worthwhile, and regardless of financial waste for patients, health care providers, insurers, and taxpayers.

Walker didn't understand this—or he pretended not to. Either way, there was no basis for a meaningful policy debate between officials sincerely interested in reproductive care.

The gap even extended to contraception access. In 2014 the U.S. Supreme Court's *Burwell, Secretary of Health and Human Services, et al. v. Hobby Lobby Stores, Inc., et al.* ruling had the effect of empowering employers to limit workers' access to birth control. Senate Democrats crafted legislation that would have required private insurers to cover birth control, just as the Affordable Care Act intended, though houses of worship would be exempt and religious nonprofits would be accommodated.

Reluctant to appear anticontraception, Senate Republicans responded with a rival proposal that would have prohibited businesses from preventing employees from purchasing birth control privately. "We plan to introduce legislation this week that says no employer can block any employee from legal access to her FDA-approved contraceptives," Senate Minority Leader Mitch McConnell said. "There's no disagreement on that fundamental point."[46]

When critics said the GOP proposal was meaningless, they didn't mean it was inadequate or misguided. As the HuffPost's Laura Bassett put it, the Republican bill "literally does nothing."[47]

In practice, if a woman's employer, taking advantage of the *Hobby Lobby* ruling, objected to covering contraception through its insurance plan, the Republican bill would have allowed her to go buy contraception with her own money. That was it. That was the whole plan: consumers making private transactions in the private marketplace, buying a product in exchange for money.

In effect, the GOP idea was to tell the public, "See? Even if your employer opposes contraception, your boss can't stop you from going to the store and paying for medicine you may want to voluntarily purchase." Or put another way, the Republican alternative to Democratic legislation was little more than a legislative reminder to the country that birth control is legal and people can buy it if they want to—and if they can afford it.

It badly missed the substantive point of the controversy created by the *Hobby Lobby* ruling. The Affordable Care Act intended to make contraception access a standard feature of coverage; conservatives on the Supreme Court said private, for-profit corporations can be excluded from meeting this standard if they see birth control as morally offensive. What did the Senate Republican bill have to do with this dilemma? Absolutely nothing.

GOP policy making on the broader issue did not improve in the years that followed. After Brett Kavanaugh replaced Anthony Kennedy on the U.S. Supreme Court, opponents of reproductive rights concluded that the door was open to overturning the 1973 *Roe v. Wade* precedent, and they began legislating accordingly.

In Alabama, for example, state GOP policy makers were responsible in 2019 for the most sweeping abortion ban in the country. During the legislative debate, a Democratic lawmaker asked one of the bill's chief sponsors, Republican state senator Clyde Chambliss, why Alabama should block rape victims from terminating an unwanted pregnancy. The Republican tried to set everyone's minds at ease.

Under the new policy, he said, "anything that's available today is still available up until that woman knows she's pregnant. So she has to take a pregnancy test, she has to do something to know whether she is pregnant or not. You can't know that immediately. It takes some time for all those chromosomes and all that that you mentioned. It doesn't happen immediately."[48]

Chambliss did not appear to be kidding. The architect of Alabama's antiabortion law publicly defended his proposal by suggesting that rape victims can still obtain abortions just so long as they don't know they're pregnant, which was every bit as bewildering as it seemed. He also appeared to confuse chromosomes and the hormones used in pregnancy tests.

Asked by a reporter if an incest victim could get an abortion in Alabama, the Republican added, "Yes, until she knows she's pregnant." Chambliss was also asked what his bill's reference to "attempted abortions" meant, and he said he wasn't altogether sure.[49]

One female legislator suggested Chambliss may not be as familiar as he should be with the reproductive process. He replied, "I don't know if I'm smart enough to be pregnant."[50] It led the *Washington Post*'s Dana Milbank to respond, "The better question is whether he's smart enough to be writing laws."[51]

After the slapdash Alabama policy passed, Donald Trump celebrated the abortion ban and accused Democrats of supporting a system that allows "a baby to be ripped from the mother's womb moments after birth."[52] It led his critics to add physiology to the list of things the president didn't understand as well as he should.

That conservative opponents of abortion rights scrambled to pass new restrictions following Kavanaugh's controversial confirmation wasn't surprising. It stood to reason that those eager to reverse *Roe* would try to take advantage of their best opportunity in nearly a half century. What was surprising was the degree to which Republican policy makers working on the issue seemed to have no idea what they were doing.

Ohio's GOP majority had already approved a series of related restrictions on reproductive rights, but in 2019 it went considerably further, taking up a bill to ban abortions after six weeks of pregnancy, which is problematic for obvious reasons: many women don't necessarily know if they're pregnant after just six weeks.

A follow-up proposal in Ohio intended to prohibit insurers from covering abortion services, irrespective of the private businesses' and customers' wishes. Many legal observers warned that the hastily written bill, if implemented, could have the effect of limiting access to commonly used forms of contraception—which, ironically, would lead to more unwanted pregnancies and more people seeking abortions.

A local news station in Ohio asked state representative John

Becker, a Republican and the author of the legislation, to respond to those concerns. "I don't know because I'm not smart enough to know what causes abortions and what doesn't," he said. "The bill is just written, 'if it causes an abortion,' and people smarter than me can figure out what that means."[53]

It was the distillation of post-policy thinking in the span of about ten seconds. The architect of antiabortion legislation in one the nation's largest states simply wanted to pass a bill in pursuit of his political goal. He hadn't thought through the substantive details, and when pressed on the ambiguities of his own plan, Becker was content to let others—those with greater policy literacy than the author of the legislation—do his homework for him.

Children in math classes have been told for many years, "Show your work." It's not enough to simply jot down an answer; teachers want to know that students have thought through the problem and worked deliberately toward an answer. As the Republican Party abandoned its status as a governing party, it also moved away from showing its work and demonstrating its capacity for working carefully toward solutions. GOP officials are too often content to simply jot down a satisfying answer, confident that, to borrow a phrase, someone smarter than them will consider the practical details later.

CHAPTER 9

"Governing by Near-Death Experience"

Government Shutdowns and Debt-Ceiling Crises

FOR MOST OF AMERICAN history, government shutdowns simply did not happen. These crises didn't begin until after policy makers approved the Congressional Budget Act of 1974, which was intended to apply discipline to the process through which lawmakers spent federal funds.

The policy shift may have been well intended, but it opened the door to ugly crises. When the U.S. government shuts down, the incidents capture an obvious breakdown in governance, but the practical effects are felt widely: the federal government—the nation's single largest employer—furloughs much of its workforce, leaving hundreds of thousands of Americans without paychecks.[1] At the same time, federal operations are disrupted, important public-sector work goes undone, and the domestic economy takes a hit.

If shutdowns were rare, the damaging nuances would be less alarming, but the breakdowns have actually become far more common as the Republican Party becomes less and less invested in governance. Indeed, government shutdowns have come to represent the logical conclusion of the GOP's post-policy posture: Republican officials, indifferent to responsible governance, and reluctant to engage in mature policy making, have begun regularly bringing the gears of government to a halt, often for reasons they've struggled to explain or defend.

There were some incidents along these lines in the 1970s, but federal officials weren't furloughed, and the brief interruptions didn't resemble what most Americans would now characterize as proper shutdowns. Even at the time, they were generally described as "funding gaps" and "budget shortfalls" that carried limited consequences for the public. For example, tourists did not arrive in the nation's capital and discover locked doors at Smithsonian museums.

The 1980s and 1990s brought an uptick[2] in the frequency of these incidents, but shutdown politics took a dramatic turn after Republicans took control of the House in 2011. It's not a coincidence that the routinization of manufactured crises coincided with the GOP's wholesale embrace of post-policy politics.

In 2013 the *Washington Post*'s Karen Tumulty described the dynamic as "governing by near-death experience," adding, "It is as though Washington has had backward evolution—operating as a primitive, leaderless village where petulance passes for governance."[3]

That intemperance came to the fore in February 2011, just one month into the new House Republican majority, when GOP leaders announced plans for a government shutdown unless the Obama White House and the Senate Democratic majority agreed to $100 billion in spending cuts.[4] As a matter of economics, the fragility of the nascent post–Great Recession recovery was quite real, and the idea of taking that much capital out of the economy, even gradually, was plainly absurd.

What's more, as a matter of policy making, Republicans hadn't exactly done their homework: the $100 billion figure was arbitrary. This wasn't a dynamic in which GOP lawmakers had identified $100 billion in wasteful government spending and expected Democrats to recognize the value of the cuts; Republicans simply picked a round number that sounded nice, confident that officials would figure out the details later.

That shutdown was narrowly averted, with the parties working out a modest deal roughly ninety minutes before the deadline. But in September 2011, Republicans threatened another shutdown,[5] only to make the same threat in April 2012.[6] What was once considered a radical and intimidating tactic, involving a histrionic political standoff, had become something GOP officials were comfortable throwing around casually once or twice a year.

In late September 2013, Republicans not only repeated the threat but also decided that saber-rattling without follow-through would no longer suffice.

The shutdown plan did not unfold overnight. Months earlier, seventeen GOP senators, led initially by Senators Mike Lee and Marco Rubio, vowed to fight against any spending bill that kept the government open that included funding for the Affordable Care Act. The idea faced some intraparty resistance.

"I think it's the dumbest idea I've ever heard of," Republican Richard Burr said months before the shutdown deadline. "Listen, as long as Barack Obama is president, the Affordable Care Act is going to be law." The North Carolina senator added that he'd served in Congress during the Republicans' nearly eight-week government shutdown beginning in mid-November 1995—an ordeal that helped Bill Clinton's political standing and did lasting harm to the reputation of Newt Gingrich, the House Speaker who orchestrated the scheme—and learned from the experience.[7]

"I think some of these guys need to understand that if you shut down the federal government, you better have a specific reason to do it that's achievable," Burr said. "Defunding the Affordable Care Act is not achievable through shutting down the federal government. At some point, you're going to open the federal government back up, and Barack Obama's going to be president, and he won't have signed [a bill to derail] the Affordable Care Act."

In the House, Republican Tom Cole of Oklahoma, a deputy majority whip and close ally of House Speaker John Boehner, added, "Seems to me there's appropriate ways to deal with the law, but shutting down the government to get your way over an unrelated piece of legislation is the political equivalent of throwing a temper tantrum. It's just not helpful."[8]

Burr and Cole offered their fellow Republicans sound advice, which the party proceeded to ignore.

The ever-combative senator Ted Cruz took over as one of the ringleaders of the shutdown brigade, presenting his GOP brethren with a simple talking point: when it came to derailing Obamacare, this was basically their last chance. Though many of the ACA's consumer protections took effect after the bill became law in 2010, as of 2014, American families would start receiving subsidized coverage through plans chosen by way of exchange marketplaces. If Republicans weren't prepared to pull out all the stops in late 2013, it would be vastly more difficult for the party to strip millions of Americans of their benefits once the law was fully implemented.

The more Cruz faced skepticism, the more the far-right Texan resorted to mocking GOP leaders for resisting his strategy and rallying grassroots activists behind the cause.[9] As the shutdown deadline drew closer, Cruz engaged in some memorable theatrics on the Senate floor, reading Dr. Seuss's *Green Eggs and Ham* and arguing that it applied to the debate over the future of the Affordable Care Act. "The difference with green eggs and ham—when Americans tried it, they discovered they did not like green eggs and ham, and they did not like Obamacare, either," the senator said. "They did not like Obamacare in a box, with a fox, in a house, or with a mouse."

Cruz probably should have read the children's book all the way to the end. If he had, the Republican might have noticed that ultimately the protagonist discovers green eggs and ham really aren't so bad after all. Indeed, the character comes to regret criticizing something he didn't fully understand and ends up celebrating the very thing he'd complained about so bitterly. Cruz said the story had "some applicability, as curious as it may sound, to the Obamacare debate." As it turned out, ACA proponents drew a similar allegorical conclusion.

The book reading fell flat, but Cruz's campaign intensified. In late September 2013, the Texan—less than one year into his congressional tenure—started working directly with House GOP members, urging rank-and-file lawmakers to ignore guidance from the House Speaker's office and instead follow Cruz's advice.[10] The tactics had the intended effect: Boehner crafted a series of proposals intended to avoid a government shutdown, and each was rejected by his own members. It's just one of the reasons that Boehner later referred to Cruz as "Lucifer in the flesh."[11]

Republican opponents of the shutdown strategy gradually fell silent. On October 1, 2013, GOP lawmakers shut down the government for the first time since Newt Gingrich's failed gamble in 1995–96.

The idea that Republicans were acting like members of a governing party was plainly laughable. They were so disgusted by the idea of working-class families gaining access to subsidized health insurance that they wrote a clumsily written hostage note to Barack Obama, telling him to stop implementation of the Affordable Care Act, or federal operations would remain closed indefinitely.

The Republican strategy had all the sophistication of a brick thrown through a window. Cruz and his cohorts seemed to believe it was entirely realistic that the Democratic president would wake up one morning and say, "Well, I want the government to open, so millions of Americans will just have to go without access to affordable medical care for a while."

The odds of this happening were nonexistent, which was a nagging detail most GOP lawmakers were well aware of. The post-policy party closed down the government anyway, lacking any kind of coherent plan for success. Nine months into the 113th Congress, Republicans were already demonstrating their inability to govern responsibly—and their unwillingness to even try.

Less than a year earlier, in the 2012 elections, a Democratic president won reelection fairly easily; voters expanded the Democrats' U.S. Senate majority; and Democratic U.S. House candidates received more votes than their GOP counterparts. The Republican majority existed largely as a result of gerrymandered congressional districts.

It was against this backdrop that the far-right House majority, unsatisfied with voters' choices, found it appropriate to launch a maximalist shutdown scheme, election results be damned.

The traditional American policy making process had worked reasonably well for generations. Lawmakers with an idea or a goal could introduce legislation. If it cleared one chamber, its proponents could work to pass it through the other. If they succeeded, the idea's supporters could try to persuade the White House about its merits. It's not an easy system, to be sure, but it's not supposed to be. The process was designed to be slow and difficult, featuring a series of choke points.

As was evident with their shutdown scheme, Republicans came to believe it was time to add a shortcut to the process. Why rely on the traditional legislative process when lawmakers could take hostages and demand their opponents embrace their ideas, irrespective of the voters' will?

On the shutdown's first day, GOP leaders dispatched eight House Republicans to a conference room, asking Democrats to join them for budget talks. The offer came with some strings attached: GOP officials said they expected to follow a process in which Republicans wouldn't make any concessions, while Democrats would have to agree to undermine some elements of the nation's health care law, at least for a while.[12]

Democrats responded that they'd be happy to have real budget negotiations right after Republicans reopened the government.

GOP officials were so pleased with the stunt that they publicized a series of photographs of Republican House members sitting across from empty chairs, expecting the public to see the images as proof of the House GOP majority trying to be reasonable. What the homogenous party didn't seem to appreciate was the visual context: the lawmakers at the table were eight middle-aged white guys who'd just shut down the government because they were afraid of uninsured Americans getting coverage.[13]

Just like Clint Eastwood and his cringeworthy appearance at the 2012 Republican National Convention, in which he had a strange, televised conversation with a stool, GOP leaders had managed to pick a fight with empty chairs and lose.[14]

As the shutdown continued, its Republican architects pitched a series of proposals they described as "compromises," each of which had one thing in common: GOP lawmakers would consider ending the shutdown only if Democrats agreed to undermine the Affordable Care Act in some material way. Republicans were willing to negotiate over the scope and scale of the damage, just so long as Obamacare was left weaker and the public benefits were lessened.

Democrats, not surprisingly, balked and seemed bewildered by their opponents' tactics. Writing for the website ThinkProgress, Judd Legum published a popular tweet summarizing the nature of the dispute:

Republicans: Can I burn down your house?
Democrats: No.
Republicans: Just the 2nd floor?
Democrats: No.
Republicans: Garage?
Democrats: No.
Republicans: Let's talk about what I can burn down.
Democrats: No.
Republicans: You aren't compromising![15]

As the pressure built and Republicans looked for a way out of the mess they'd created, some members struggled to articulate the point of their own party's strategy. Representative Marlin Stutzman of Indiana said of the shutdown, "We're not going to be disrespected. We have to get something out of this.

"And I don't know what that even is."[16]

It was a painful example of post-policy politics run amok. The Republican Hoosier and his party had shut down the government based on a strategy that didn't make sense, and to hear Stutzman tell it, the GOP members expected some kind of reward for their trouble—even if some Republicans weren't quite sure what kind of ransom would suffice.

To the extent that the party had an unobtainable policy goal in mind, its vision on how to reach the destination, or perhaps look for a new one, had grown murky. Congressman Michael Grimm of New York, before having to resign in disgrace and serve an eight-month prison sentence for tax evasion, added, "This is not just about Obamacare anymore."[17]

What *was* it about? No one on the Republican side seemed able to say. Strategy and governing had been replaced by primal instincts and a partisan id.

During the shutdown's first week, Senate Republicans met privately for a weekly gathering, where Cruz was pressed for some kind of explanation, which he could not provide. "It was very evident to everyone in the room that Cruz doesn't have a strategy—he never had a strategy, and could never answer a question about what the endgame was," one GOP senator told Politico.[18]

Reflecting on the party's rampant incompetence, a GOP staffer described House Republicans to the *Washington Examiner* this way: "They are a majority party that wants to be a minority party."[19]

It was a poignant observation. Minority parties, practically by definition, are freed from the burdens of responsibility because its members lack power. Minority parties can devote the bulk of their time to criticizing from the sidelines, while policy makers in the majority grapple with the arduous and unglamorous work of trying to solve problems.

Republicans found comfort playing the role of minority party, because the status was easier than having to produce meaningful results. Once placed in the role of a governing party, however, GOP officials were lost and directionless, unable and unwilling to craft policies, while stumbling into a shutdown debacle of their own making for reasons they couldn't explain. The result was a case study in the collapse of Republican governance.

Seventeen days after the House GOP hatched its shutdown scheme, party leaders grudgingly brought the fiasco to an end, advancing the same spending bill they'd refused to pass weeks earlier.

It was an excruciating process to watch, but it wasn't the only example of Republicans pushing a manufactured crisis in the Obama era. In fact, the 2013 shutdown wasn't even the most damaging of the self-imposed calamities.

Under the American system, Congress has the power of the purse and appropriates federal funds, but it's the executive branch that is responsible for actually spending the money. In theory, there's nothing wrong with the model. In practice, however, policy makers routinely run into a glitch in the process known as the debt ceiling.

The executive branch spends taxpayers' money and covers the country's financial obligations, but it lacks the legal authority to spend more than the country takes in. Since the United States nearly always runs an annual budget deficit, this means administrations have spent decades going back to Congress and asking members to extend the nation's borrowing limit—in effect, getting permission to spend all of the money lawmakers already allocated.

It's a straightforward process that doesn't cost anything; raising the ceiling simply allows the country to cover the debts it's already accrued. Failure to do so would result in the United States defaulting on its debts, sparking a crisis that would shake the global economy.

Across most of the twentieth century, the process of raising the debt ceiling routinely got a little messy—lawmakers have used the process for grandstanding on multiple occasions—but neither party had ever seriously entertained the possibility of trashing the full faith and credit of the United States government.

That is, before House Republicans claimed their majority status in January 2011.

Shortly after the 2010 midterms, when GOP candidates made massive gains in Congress's lower chamber, incoming House Speaker John Boehner realized that he and his party would have to raise the debt limit, despite Republicans' recent rhetoric condemning the practice. The GOP leader told reporters two weeks after the elections, "I've made it pretty clear to [the incoming Republican majority] that as we get into next year, it's pretty clear that Congress is going to have to deal with this. We're going to have to deal with it as adults. Whether we like it or not, the federal government has obligations, and we have obligations on our part."

He added soon after, "We'll have to find a way to help educate members and help people understand the serious problem that would exist if we didn't do it."[20]

Almost immediately, far-right members, riding a Tea Party wave, suggested they weren't at all prepared to address the nation's borrowing limit "as adults." Ohio representative-elect Bill Johnson said in late November 2010, "Most of us agreed that to increase the limit would be a betrayal of what we told voters we would do."[21] Whether he knew exactly what the debt ceiling was wasn't altogether clear.

But it didn't appear to matter. In the weeks and months that followed, Republican lawmakers in both chambers practically tripped over one another, scrambling to assure the party's base that they too would oppose paying the nation's debts. In January

2011 Senator Lindsey Graham appeared on NBC's *Meet the Press* and emphasized that failing to raise the debt ceiling "would be very bad" for the country.[22] In the same interview, the South Carolinian added that he was nevertheless prepared to go along with his party's plans to hold the debt ceiling hostage.

History offers examples of struggling countries, facing financial turmoil and domestic upheaval, defaulting on their debts, but what Republicans were prepared to do in 2011 was something altogether new: the United States confronted the possibility of default, not because the country lacked the resources needed to meet its obligations, but because lawmakers from one party simply didn't want to write the check for the things the country had already bought.

Austan Goolsbee, chairman of Barack Obama's White House Council of Economic Advisers, appeared on ABC's *This Week* in January 2011 and tried to explain the circumstances. "The debt ceiling is not something to toy with," the economist said, adding, "If we get to the point where you've damaged the full faith and credit of the United States, that would be the first default in history caused purely by insanity. . . . We shouldn't even be discussing that. People will get the wrong idea. The United States is not in danger of default."[23]

A month later, Boehner shared a similar sentiment but arrived at a different conclusion. The new House Speaker told Fox News that failing to raise the debt ceiling would be "a financial disaster" for the United States and the world. Then he added that he and his party would nevertheless refuse to act responsibly unless Democrats met Republican demands for "significant" cuts to federal spending.[24]

Boehner didn't specify what that might mean—by all appearances, he had no idea what cuts would satisfy his party—and neither did Senate Minority Leader Mitch McConnell, who said in March 2011 that his caucus was also prepared to kill a debt limit extension unless it included "significant" steps to reduce the nation's deficit.[25]

The party didn't have a plan, per se, but Republican leaders had something they liked more. As Senator John Cornyn of Texas put it in a tweet, GOP officials saw the debt ceiling as "the ultimate leverage."[26] It was a posture lacking in any sense of subtlety: a major American political party signaling a willingness to crash its own country's economy, on purpose, unless its demands were met.

The politically violent nature of the threat was itself extraordinary, but there was also a degree of hypocrisy to the tactics. The GOP officials demanding a focus on debt reduction were the same Republicans who'd spent the Bush-Cheney era adding trillions of dollars to the debt. After ignoring fiscal concerns for nearly a decade, Boehner, McConnell, and their cohorts suddenly decided—once a Democrat was in the Oval Office—that reducing the deficits they'd created was the only goal that mattered.

If that meant taking money out of a fragile economic recovery, so be it. If that meant threatening a catastrophic default on the nation's debts, that too was a price Republicans were prepared to pay. (That GOP lawmakers would adopt the opposite posture during Donald Trump's presidency only added insult to injury.)

There were no comparable circumstances in the American tradition. A major political party, en masse, hadn't threatened the United States with such widespread harm since the Civil War.

By mid-April 2011, leaders from the American business community, after spending a few months assuming that Republicans were engaged in a pointless and inconsequential stunt, started feeling antsy. Groups such as the U.S. Chamber of Commerce and the National Association of Manufacturers, among others, told GOP lawmakers—the organizations' longtime allies—that it was time to stop screwing around.[27]

Soon after, Boehner reportedly started having conversations with top Wall Street executives, asking how long he and his party could keep the crisis going without destroying the burgeoning recovery. According to a Politico report, the executives said that "even pushing close to the deadline—or talking about it—could have grave consequences in the marketplace."[28]

One executive added, in reference to congressional Republicans, "They don't seem to understand that you can't put everything back in the box. Once that fear of default is in the markets, it doesn't just go away."

The GOP House Speaker pressed on anyway, without a plan and without any real regard for consequences.

As the deadline drew closer, bipartisan negotiations began, though the parties were playing by two very different sets of rules. Democrats from the Obama White House and Capitol Hill thought policy makers were trying to negotiate a long-term fiscal deal that would bring the budget closer to balance. In effect, the party told its GOP colleagues, "We all want to get the nation's fiscal house in order, so let's work something out."

Republican leaders, however, saw the circumstances through very different eyes. In late June 2011, during a meeting among congressional leaders, Democrats suggested an agreement that reduced the deficit through a combination of spending cuts and increased revenue. Republicans not only balked but also pushed back against the idea that the discussion was itself a negotiation.[29]

As far as the GOP was concerned, those who've been handed a ransom note are not supposed to seek concessions from the hostage takers. The party saw value in circumventing traditional American governance and achieving unnecessary goals through threats of national harm. Republicans weren't trying to govern; they were trying to extort.

By the first week in July, David Brooks, a center-right columnist for the *New York Times,* pointed to the debt ceiling crisis as evidence that Republicans might not be "fit to govern."[30] A week later, Boehner helped prove Brooks right, saying in reference to the president, "This debt limit increase is his problem."[31] The Speaker who'd stated eight months earlier that he and his party would have no choice but to deal with the debt ceiling "as adults" had transformed, adopting the rhetoric of his party's most radical crackpots.

It was around this time that the details of the Republican hostage note took shape. House GOP leaders expected, among other things, cuts to Social Security and Medicare, as well as the repeal of the health coverage expansions in the Affordable Care Act.[32]

Boehner soon after delivered a nationally televised address in which he told the public that if Obama would only agree to Republican demands, "the crisis atmosphere he has created will simply disappear." He did not appear to be kidding about who bore responsibility for the ongoing disaster.

The crisis was ultimately resolved in early August, but not before extensive damage was done to the domestic economy, the United States' reputation, and Americans' wallets. (A report from the Congressional Budget Office found that the entire fiasco ended up costing American taxpayers $1.3 billion.[33])

The day before Obama signed the bill ending the debacle, Mitch McConnell sat down with Fox News's Neil Cavuto and looked ahead to the future. The debt ceiling crisis "set the template for the future," the Senate GOP leader boasted. "In the future, Neil, no president—in the near future, maybe in the distant future—is going to be able to get the debt ceiling increased without a reignition of the same discussion of how do we cut spending and get America headed in the right direction. I expect the next president, whoever that is, is going to be asking us to raise the debt ceiling again in 2013, so we'll be doing it all over."[34]

McConnell told the *Washington Post* a day later, in reference to the statutory debt limit, "What we did learn is this: it's a hostage that's worth ransoming."[35]

There is a bare-minimum standard to public service: those put in positions of public trust should never be prepared to deliberately hurt the people they represent. As Republicans moved further and further away their status as a governing party, their debt ceiling crisis offered evidence that it was a standard they failed to meet.

There was never even a policy rationale behind the GOP-imposed tragedy. By 2011, the budget deficit Obama inherited from his Republican predecessor was already starting to shrink, and the national debt was in no way adversely affecting the economy. All Congress had to do was extend the nation's borrowing authority—a simple procedural move that costs nothing—to cover spending lawmakers had already approved.

There was a lingering jobs crisis in the United States at the time—during the GOP's debt ceiling scheme, the nation's unemployment rate was 9 percent—but Republicans ignored it entirely, turning their attention instead to a hostage standoff over a problem that did not exist.

It wasn't the first time that Americans witnessed scorched-earth political tactics in the nation's capital, but it was the first time they saw these tactics deployed for no good reason. After the 2012 election, which Barack Obama won by a comfortable margin over Mitt Romney, it came time again to raise the debt ceiling, and Mitch McConnell, true to his word, signaled an intention to hold the debt ceiling hostage, expecting Democrats to accept deep cuts to social insurance programs. In fact, in January 2013 the Kentucky senator wrote an op-ed urging Democrats to act quickly, "rather than waiting until the last minute."[36]

McConnell wasn't content to threaten the country's welfare again; he began the process by complaining about the speed with which Democrats might meet his demands.

But it didn't work. Obama, fresh from his reelection victory, told GOP leaders in no uncertain terms that he would no longer negotiate with Americans who threatened to harm the country on purpose. Republicans spent months testing his resolve, but the Democrat did not budge.

In one especially striking instance, ahead of a debt limit vote in the fall of 2013, House GOP members put together a debt ceiling hostage note—more like a Christmas gift list for Santa—in which they proposed that Democrats give Republicans the following:

delayed implementation of the ACA, the Keystone XL pipeline, Medicare cuts for wealthier beneficiaries, medical-liability reforms, changes to the Dodd-Frank Wall Street Reform and Consumer Protection Act enacted only three years before, increased oil drilling, *and* an end to the EPA's efforts to combat the climate crisis.[37] In exchange, Republicans would give Democrats nothing—except a debt ceiling increase, which Congress had to do anyway.

After the proposal leaked, Erza Klein wrote in the *Washington Post,* "John Boehner isn't even trying to pretend his House of Representatives is a sane place anymore. The House GOP's debt limit bill . . . isn't a serious governing document. It's not even a plausible opening bid. It's a cry for help."[38]

Democrats treated the Republicans' hostage note as the joke it was and made clear the party would pay no ransoms going forward.

There were a handful of related, sporadic threats during Obama's second term, but no shutdowns or debt ceiling crises for the remainder of the Democrat's presidency.

Dan Pfeiffer, a senior adviser to the president, told CNN that the Obama White House "is for cutting spending. We're for reforming our tax code, for reforming entitlements. What we're not for is negotiating with people with a bomb strapped to their chest."[39] If Republicans wanted to engage in fiscal talks, they'd have to rely on the traditional American policy-making process.

By the time Donald Trump was elected, GOP officials had largely stopped pretending to care about fiscal issues at all. That did not mean, however, that the party's interest in manufactured crises ended.

On the contrary, they began anew for reasons that were equally unwise. Just three months into his term, Trump showed great interest in the public relations significance of his hundred-day mark, which led the new president to send word to Capitol Hill that he was prepared to shut down the government unless lawmakers approved funding for a giant concrete wall that would sit along the U.S.-Mexico border.[40] His campaign rhetoric about forcing America's southern neighbors to finance construction of the project had already been put aside. Trump now expected U.S. taxpayers to pick up the tab.

Mick Mulvaney, the far-right White House budget director, suggested he didn't much care whether or not the government stayed open, telling CNBC in April 2017, "I think the consequences [of government shutdowns] have been blown out of proportion."[41] This was consistent with what we know about Mulvaney's approach to governing: while in Congress, the Republican, a founder of the right-wing House Freedom Caucus, inexplicably celebrated the 2013 shutdown as "good policy."[42]

The budget chief added that "elections have consequences," and it was the White House's expectation that wall funding would part of the spending bill that kept the government's lights on.

Trump and his team had stumbled into this without a plan or any real forethought, and they seemed wholly unaware of the fact that they'd backed themselves into a corner for no reason. If the president followed through on his threat, he'd shut down the government over an unpopular demand just as his first term was getting under way. If Trump backed off, he'd signal to Congress and the world that he was a paper tiger.

Four days later, the president blinked and told Congress his threats were just a passing whim. Treasury Secretary Steven Mnuchin reassured the public. "The president is working hard to keep the government open." In this case, by "working hard," he apparently meant "crawling away from the corner he backed himself into without any plan for success."[43]

Writing for the *New Republic,* Alex Shephard explained, "Trump could not have screwed this up more badly. He spent a week of valuable time threatening to derail a rarity in contemporary Washington: a fairly reasonable negotiation between the two parties. And he didn't get anything out of it. If anything, he showed once again that he's all bluster—that he'll make a lot of noise but, when push comes to shove, ultimately back down."[44]

Less than a year later, it was again time to prevent a shutdown, and as the deadline approached, the president announced the end of the popular Deferred Action for Childhood Arrivals program, established to extend protections to the young immigrants known as Dreamers, who were brought to the United States illegally as children. Congressional Democrats wanted to negotiate a permanent solution to the problem Trump had created, and they tied the effort to the funding bill. Complicating matters, Republicans had allowed federal funding for the Children's Health Insurance Program (CHIP) to lapse, putting at risk the program's low-income beneficiaries, and Democrats expected the spending package to address this problem, too.

A variety of solutions was considered, each of which was rejected by the White House. Senate Republicans needed Democratic votes to pass a spending bill to keep the government in business, and when those votes weren't there, the government shut down on January 20, 2018—the first anniversary of Trump's inauguration. It was the first proper shutdown in American history in which one party controlled the White House, the U.S. Senate, and the U.S. House.

According to GOP leaders, the breakdown was Democrats' fault, since they brought the immigration and children's health care issues into the fight over the spending bill. But as Jamelle Bouie explained in a piece for Slate, the argument had it backward: Democratic efforts were the predictable reaction to Republicans' failure to govern responsibly.[45] The party was in "an emergency of its own making," Bouie wrote, adding that the shutdown was an extension of the GOP's refusal "to act as a governing party."

In May 2017 Trump wrote on Twitter that he believed the United States needed a "good" shutdown.[46] In January 2018, he had one, thanks in large part to his incompetence.

During the previous shutdown, five years earlier, the future Republican president said he knew exactly how to prevent these crises from occurring. "Problems start from the top. They have to get solved from the top," Trump said in 2013. "The president's the leader, and he's got to get everybody in a room, and he's got to lead. . . . The right guy would get everybody into a room and would make a deal. You gotta get 'em into a room. You gotta to talk to them. You gotta cajole. You gotta do what you do when you make deals."[47]

Five years later, Trump was in the Oval Office, where he failed by his own standards, struggling to even try to prevent a shutdown. Presidential historian Jon Meacham, reflecting on the developments, said, "This is what government would look like without a president."[48]

Democratic senator Brian Schatz of Hawaii added, "I've never seen such a flawed negotiation. No one is in charge. . . . It's as bad as it looks."[49]

The shutdown was a mercifully brief three-day affair, resolved after the deal maker in chief extricated himself from the process. Once the shutdown was under way, the *Washington Post* reported that Trump preferred a "hide-and-tweet strategy," which White House officials approved of because it meant the president wouldn't work on an agreement his far-right base might dislike.[50] CNN added that the president told congressional leaders on the shutdown's second day that they should work out a deal on their own and present it to him once it was done—as opposed to Trump taking a hands-on role in the negotiations.[51] The key to cleaning up the mess was for the sitting president to do nothing except get out of the way.

In the words of the *New York Times,* even if Trump had been inclined to engage in negotiations, he wouldn't have been able to contribute much to the policy discussions, since he had very little understanding of the debate and was uninterested in getting up to speed. The president was "either unwilling or unable to articulate" the policies he wanted, "much less understand the nuances of what it would involve."[52]

The article went to note that Trump, having created the conditions that led to the shutdown, was "unusually disengaged," leaving the process in others' hands.[53] The post-policy party's post-policy president was content playing the role of bystander.

As a candidate, Trump declared with pride in his Republican National Convention speech, "Nobody knows the system better than me, which is why I alone can fix it."[54] But during his first shutdown, it became obvious that no one in a position of authority knew less about the system than Trump, and he alone couldn't fix anything.

As 2018 drew to a close, congressional leaders from both parties were cautiously optimistic that they could avoid another shutdown. Trump had other ideas. In a televised Oval Office meeting in December 2018, the president declared to Democratic leaders, "I am proud to shut down the government for border security. . . . I will take the mantle. I will be the one to shut it down. I'm not going to blame you for it."[55]

The message was the opposite of the one Republicans wanted to hear. Not only had Trump signaled his intention to shut down the government over border wall funding lawmakers would not approve, but the president said on camera that he would personally bear responsibility for the breakdown.

Trump was nevertheless delighted, telling House Speaker Paul Ryan the next morning that the television ratings for the Oval Office meeting had been "great."[56] The boast was bizarre: viewership data from the previous afternoon weren't available yet, and those who had tuned in saw the president make counterproductive comments his party didn't want to hear.

But true to form, Trump didn't much care about the substance of what was said during the talks. What mattered was whether he believed people had watched in significant numbers. The president's priority was capturing the spotlight, even if it was unflattering. In the process, Trump drew a line in the sand: either Congress would approve $5.6 billion in taxpayer funds for a giant border wall or he would shut down the government a few days before Christmas. His congressional allies wanted to see a smarter approach, but they had no idea how to change the president's mind.[57] Vice President Mike Pence told Democrats the president would accept $2.5 billion, but Trump publicly contradicted him soon after, reinforcing concerns about rampant White House dysfunction.[58]

Senate Majority Whip John Cornyn, whose job included wrangling GOP votes, conceded a few days before the deadline that if there was a plan to avoid a shutdown, "I'm not aware of it."[59]

With hours remaining before the shutdown, Trump assured the public he was "negotiating with the Democrats" to resolve the standoff. No one knew what he was talking about: the president hosted a White House luncheon with far-right GOP lawmakers, but Democrats weren't invited.[60] It wasn't clear what he would have told Democrats even if the claim were true.

Even once the shutdown was under way, Trump struggled to discuss the developments in ways that suggested he understood what was going on around him. On Christmas Day, for example, the president bragged with great enthusiasm about government contracts he'd awarded for the construction of border barriers. Those contracts did not exist in reality, and whether Trump knew this or not, in the American system of government, a president can't unilaterally give out federal contracts at his own discretion.

A day later, the Republican claimed to have heard from "many" federal workers, furloughed during the shutdown, who told him they'd rather have a partially built border wall than a paycheck.[61] There was no evidence those exchanges occurred outside of Trump's imagination.

The chief executive stayed in the White House over the winter holidays, but at no point did he reach out to Democratic leaders about finding a way out of the mess he'd created.[62] If his tweets were any indication, the president watched a staggering amount of television but did not appear to do any substantive work at all on his own political scheme.

He whined to Fox News's Sean Hannity on the air, "I could have enjoyed myself. I haven't left the White House because I'm waiting for them to come over in a long time. You know that. I stayed home for Christmas. I stayed at the White House for New Year's. . . . My family, I told them, 'Stay in Florida and enjoy yourselves.' The fact is, I want to be in Washington. I mean, I consider it very, very important."[63]

Except, if Trump genuinely believed that his not leaving the White House over the holidays was "very important," he would have taken steps to bring the shutdown to a close—or at least tried to learn what might help bring about such a goal. The president didn't do anything of the kind, which made it all the more curious that he sought pity during a nationally televised interview. After all, Trump was the one who'd created the standoff in the first place, causing real hardships for people who weren't hanging out in a presidential mansion.

At a cabinet meeting on January 2, 2019, Trump went on to tell his team, "I was in the White House all by myself for six, seven days, it was very lonely. My family was down in Florida. . . . But I felt I should be here just in case people wanted to come to negotiate the border security." Later during the meeting, he added, "I was hoping that maybe somebody would come back and negotiate, but they didn't do that, and that's okay."

They were the unscripted comments of a man who didn't know enough about the process to realize how pathetic his perspective was. To hear Trump tell it, his plan to resolve the government shutdown was to sit around for days, feeling lonely and waiting to see if someone stopped by to negotiate with him.

Nearly two weeks into the shutdown, Trump's focus was less on finding a solution and more on branding. The president told lawmakers he preferred the word "strike" to "shutdown," despite the definitions of the two words. He told reporters later in the day, "I don't call it a shutdown."[64] When that pitch didn't gain traction, Trump switched gears, abandoned his demand for a great big border wall, and announced that he wanted instead a series of steel-slat border barriers. The president asserted that the Democratic leadership "feels better about" moving away from concrete. He added, "As far as concrete, I said I was going to build a wall. I never said, 'I'm going to build a concrete . . .' I said I'm going to build a wall."

No one had any idea what he was talking about. For one thing, Democratic lawmakers didn't care about specific choices in building materials, and for another, Trump had actually spent several years promising Americans he'd build a border wall made of concrete.[65] It appeared at that point in the process that the president's plan to end his shutdown involved gaslighting those following the developments.

When that didn't work, either, Trump invited congressional leaders to the White House, walked into the meeting room, and asked the new/old House Speaker, Nancy Pelosi of California (who'd held the position second in line to the presidency from 2007 through 2010), if she was prepared to pay his border-barrier ransom. She said she was not, at which point the president abruptly left the gathering.[66]

The Republican's tantrum generated another round of unflattering headlines, but it also reinforced an awkward truth: Trump didn't know what he was doing. His plan involved making an unreasonable demand, untethered to credible policy goals, and waiting for skeptical lawmakers to give him what he asked for. The president knew he wanted a win, but he hadn't the foggiest idea how to get one before the shutdown, and he seemed even more lost during it.

Presented with a governing test of his own making, Trump couldn't overcome his post-policy amateurishness. The chief executive knew how to shut down the government, but when it came to every other relevant detail needed to reach an agreement—

understanding the facts, telling the truth, remaining consistent—he simply wasn't up to the job.

On the shutdown's twenty-fourth day, Trump assured the public via Twitter, "I do have a plan on the Shutdown."[67] If the plan existed, he never told anyone what it was or made any attempt to implement it.

Five days later, Trump brought members of the House Problem Solvers Caucus—a bipartisan group of relative moderates—to the West Wing for a chat. If his goal was to demonstrate a degree of competence and control, it backfired: the president ended up talking about his vision on immigration and border policies in a way that made clear he still hadn't done his homework.

Congressman Vicente Gonzalez Jr., straining to be polite, said he was struck by the president's "very serious misconceptions of the border." The Texas Democrat added, "I was listening to him today. He makes a lot of comments that are so untrue. But I believe that he actually believes them."[68]

It was an unsettling peek behind the curtain. Some observers assumed that Trump was peddling nonsense to the American people for purely political reasons, cynically trying to exploit public ignorance. A more discouraging truth was coming into focus: the president's confusion about his plan and its rationale was genuine.

A week later, Commerce Secretary Wilbur Ross caused a stir following a CNBC interview in which he said he didn't "quite understand" why some federal workers affected by the government shutdown had turned to food banks for assistance. Asked for his reaction, the president said he understood what the cabinet secretary was trying to say.

"Local people know who they are when [furloughed federal workers] go for groceries and everything else," Trump claimed, adding, "And that's what happens in times like this. They know the people, they been dealing with them for years, and they work along. The grocery store—and I think that's probably what Wilbur Ross meant."

As far as the president was concerned, he understood the lifestyles of the Americans who were no longer receiving paychecks because of the shutdown. In his vision, these workers lived in small enough communities that they could go to a local grocer who knew and cared about them and who would gladly let them buy groceries on credit. This, according to Trump, is "what happens."

Writing for *New York*, Sarah Jones marveled at just how ridiculous the president's understanding of these workers' circumstances was during the shutdown: "It is not true . . . that hungry workers in need of a gallon of milk can show up at the local Foodtown and just promise to pay the store back later. We do not inhabit the world of *Little House on the Prairie*. Half Pint cannot go to the general store and place a dozen eggs on store credit until Pa's farm starts to make money. If that were possible, workers would have taken advantage of this system already."[69]

Many GOP lawmakers were nevertheless content to play along as if their president were competent. Lindsey Graham, a Trump loyalist, said the president couldn't back down because if he did, it would probably be "the end of his presidency."[70] Senator John Kennedy added that there was no way the president would reopen the government to allow for congressional border security talks.

"You know when that's going to happen? When you look outside your window and see donkeys fly," the Louisiana Republican predicted. "It's not going to happen. . . . You can have your own opinions about President Trump, but I think most fair-minded people would have to agree he's a smart man."[71]

On January 25, 2019, Trump caved, agreeing to sign a stopgap measure that ended his thirty-five-day shutdown—the longest in the nation's history. The Republican's humiliating defeat was complete: the bill Trump embraced to reopen the government gave him nothing. Flying donkeys were nowhere in sight.

When lawmakers got to work on a new spending package with border security elements, the president seemed wholly unconcerned with its contents. In mid-February 2019, as a deal in

Congress took shape, Trump held a rally in Texas, after which he appeared on Fox News to talk about the size of his crowd. When host Laura Ingraham asked about the congressional agreement, the president conceded he didn't know what it was.

He had a choice: learning about the border security agreement or appearing on the cable news network. "I had to choose you," Trump told Ingraham, prioritizing television over governing and calling fresh attention to his post-policy attitudes.[72]

Had he been briefed on the final deal, Trump would have learned that Congress had agreed to a package that invested $1.375 billion in border barrier enhancements. Before the shutdown, lawmakers had offered the White House a bipartisan deal that would have spent $1.6 billion on the same priority.[73]

Whether the president understood this or not, by shutting down the government for thirty-five days, Trump ended up with less of what he said he wanted.

CHAPTER 10

"It's Like These Guys Take Pride in Being Ignorant"

The Eternal Campaign

THREE MONTHS BEFORE THE 2008 presidential election, Barack Obama held a wide-ranging discussion with voters in Missouri, and the event touched on energy policy and the climate crisis. After the Democrat highlighted a series of elements of his energy plan, a voter asked what a typical person could do to address the problem as an individual. The senator talked about energy conservation and briefly noted modest steps people could take to help make a difference.

Obama specifically pointed to, among other things, car owners bringing their cars in for regular tune-ups and keeping their tires properly inflated. Relying on data from the Bush administration's Energy Department, the candidate added that the amount of energy saved though routine auto maintenance would be comparable to the savings the United States would see from coastal oil drilling.

It was wholly unremarkable, except to Republicans. Soon afterward, John McCain said of his presidential rival's energy plan,

"It seems to me the only thing [Obama] wants us to do is inflate tires" to improve gas mileage. Newt Gingrich ran to Fox News to describe the Democrat's comments as "loony tunes."[1]

The Republican National Committee thought it'd be clever to create tire gauges featuring the words "Obama's Energy Plan" and deliver them to DC newsrooms. In early August 2008 Mark Salter, one of McCain's top aides, told reporters traveling with the Republican candidate that after takeoff, he'd distribute copies of the Obama energy plan. Soon after, Salter, known for a gruff demeanor, was reportedly "giddy" and "gleeful" while handing out tire gauges.[2]

Republicans weren't just acting like children with a new toy; they also exploited what they saw as a financial benefit. That same week, the McCain campaign started a fund-raising effort based on the mind-numbing story, telling donors, "Today I'm asking for your help in putting Senator Obama's 'tire gauge' energy policy to the test. With an immediate donation of $25 or more, we will send you an 'Obama Energy Plan' tire pressure gauge."

The Democratic contender had a comprehensive and detailed energy plan, as Team McCain and its allies knew. Obama was also utterly correct on the substance about added efficiency benefits from routine car maintenance. But Republicans thought there might be some political value in seizing on this in the most cynical, post-policy way possible, turning a serious issue into a punch line.

After a week of the GOP focusing aggressively on this issue, Obama could hardly contain his irritation with his rivals' juvenile antics. The future president told an audience in Ohio that Republicans "know they are lying about what my energy plan is," adding, "They are making fun of a step that every expert says would absolutely reduce our oil consumption by three to four percent. It's like these guys take pride in being ignorant. They think it is funny that they are making fun of something that is actually true. They need to do their homework, because this is serious business."[3]

Obama's unscripted frustrations were understandable. The tire gauge nonsense was hardly the first silly gimmick ever attempted by a presidential campaign, but it offered a striking example of Republicans' wholesale indifference to the substance of one of the world's most pressing issues. The Democratic candidate desperately wanted to have a real debate over a vital issue, only to find a rival campaign that literally preferred to play with toys.

Republicans' indifference to the substance of governing is routinely evident after Election Day, but it's equally apparent before ballots are cast. For much of the last decade, GOP candidates, up and down the ballot, have run on alarmingly thin platforms, published hollow "issues" pages on their websites, downplayed the importance of substantive campaigns, and demonstrated an unhealthy lack of interest in telling voters what they'd do with power once in office.

In July 2010 House Minority Leader John Boehner, eager to become House Speaker, started fielding questions from the press about what voters could expect from GOP lawmakers if they reclaimed the majority—a subject he was loath to discuss in any detail at all.[4]

The Ohioan endorsed budget cuts, but he wouldn't say which investments would face the knife. Boehner said he wanted to pass a health care plan, but he wouldn't say what kind. He looked ahead to "an adult conversation with the American people" about scaling back social insurance programs—sometimes referred to as "entitlements"—but when asked about the kind of changes he had in mind for Social Security, the Republican leader replied, "I have no idea."

Two weeks later, longtime congressman Peter King dismissed the idea of presenting voters with a meaningful agenda because, as he put it, "Then we would have the national mainstream media jumping on every point trying to make that a campaign issue."[5] It was a curious argument. The New York Republican believed his party could give the electorate a detailed sense of how his party would govern, but GOP officials and candidates would then run the risk of having to discuss those issues in detail with voters before Election Day. It was a dynamic King thought best avoided.

Except in a democracy, governing parties aren't supposed to approach the political process this way. In the Platonic ideal, candidates are supposed to want their agendas to be campaign issues, since that's the intended point of campaigns. Victors not only get the satisfaction that comes with an electoral endorsement, but also they get to credibly claim a popular mandate for their platforms. For a party to deliberately hide its ideas for fear of examination is cowardice and an implicit declaration that campaigns are supposed to be something other than a battle of substantive ideas.

As a practical matter, the motivation behind Boehner's and King's reticence was hardly a secret: their party was poised to make dramatic gains in the 2010 midterms, and the party's agenda featured a long list of priorities mainstream voters would not like. But that wasn't much of an excuse. A party that expects to lose if its agenda is exposed to sunlight has a decision to make: change that agenda or work to keep its ideas in the shadows.

Governing parties don't choose the latter. In 2010 Republicans didn't care.

Utah Senator Bob Bennett, who lost his reelection bid when GOP primary voters in his own state decided he wasn't a sufficiently rigid ideologue, reflected on his party's vision as his career came to an involuntary end. "As I look out at the political landscape now," the outgoing senator said in the summer of 2010, "I find plenty of slogans on the Republican side, but not very many ideas."[6]

Soon after, Ron Johnson, running for the U.S. Senate in Wisconsin, broke with his usual habit of avoiding journalists and agreed to sit down with the editorial board of the *Green Bay Press Gazette*. One editor noted that the GOP candidate hadn't released anything resembling a jobs plan other than repeating a vow to "cut spending," and she asked if Wisconsin voters could expect to see a "real jobs plan" from his campaign before Election Day.

Johnson replied, "Bring fiscal discipline to the federal government. You know, we've got to curb spending."

The editor, understandably confused, asked, "So, your jobs plan is to control spending. But what about the middle class?" Johnson shrugged his shoulders and said, "We have to get the economy moving."[7] It was as if the far-right candidate heard Bob Bennett's comments about his party's preference for slogans over ideas and went out of his way to prove Bennett right.

That same week, a reporter asked the Wisconsin Republican about the U.S. Department of Veterans Affairs and its responsibility toward homeless veterans. "I don't have all the details," Johnson replied. "One thing I will point out: I don't believe this election really is about details. It just isn't."[8]

His figure-it-out-later approach to governance didn't hurt his candidacy—Johnson won by 5 points—and he wasn't the only GOP candidate who adopted the post-policy posture toward campaigning. Around the same time in 2010, Linda McMahon, a wealthy businesswoman who led the company that became World Wrestling Entertainment (WWE), was running for the U.S. Senate in Connecticut, and was asked about how she'd address issues such as Medicare and Social Security if elected. The Republican replied, "I can certainly tell you I'm not adverse to talking in the right time or forum about what we need to do relative to our entitlements. I mean, Social Security is going to go bankrupt. Clearly, we have to strengthen that. . . . I just don't believe that the campaign trail is the right place to talk about that."

It was rare to hear a Republican endorse the vote-first, answer-questions-later approach to politics quite so explicitly. (McMahon, in her second attempt at a U.S. Senate seat, in 2012, suffered a double-digit defeat, but she joined Donald Trump's cabinet in 2017.)

About six weeks before the 2010 midterm elections, GOP leaders, eager to give the appearance of having some ideas, unveiled a vague and contradictory *A Pledge to America* that managed to make the Republican Party's vision even murkier. Despite the Republicans' emphasis on new ideas, host Jon Stewart of Comedy Central's *The Daily Show* aired a brutal segment noting that the party was recycling decades-old goals, in some cases, literally word for word.[9]

Worse, when John Boehner hit the Sunday shows to defend the "pledge," he didn't even try to tie the fuzzy document to the GOP's practical goals. "Let's not get to the potential solutions," the Ohioan said on *Fox News Sunday*, seemingly unaware of the fact that his party's platform was supposed to be about potential solutions.[10]

The week before the election, in a competitive Pennsylvania congressional district, Republican Mike Kelly participated in a debate and was asked to provide voters with some specific areas of the budget he intended to cut. "Sure, I'll address that," the candidate said, "and I'll address it very specifically.

"Absolutely, there's stuff to be cut. What is it right now? I can't tell you. . . . Specifically, what I would do? I would be the most responsible legislator who's out there."[11]

In 1993 NBC aired a classic episode of the beloved TV sitcom *Cheers* featuring a city councilman who went to the bar to ask patrons for their support. "Kevin Fogarty, city council. I hope I have your vote on Election Day," the candidate said. This led Dr. Frasier Crane, played by Kelsey Grammer, to ask, "And why exactly should I vote for you, Mr. Fogarty?"

"Well, because I'm a hard worker, and I take a stand."

Crane pressed on, asking, "On what, exactly?"

"The issues of the day," Fogarty replied.

"*Which are?*" Crane inquired.

"The things that concern you and your family—the most," the councilman concluded.

It was hard not to wonder whether Pennsylvania's Mike Kelly had seen the episode.

At the midway point between the 2010 and 2012 elections, as the Republican Party's presidential field started to take shape, Ed Rogers, a prominent Republican pollster, wrote a piece for the *Washington Post* sharing his impressions of the party's base. GOP voters, he argued, "don't have much tolerance for too many facts or too much information. In politics, a bumper sticker always beats an essay."[12]

Rogers's candor was welcome—most of the leading voices in Republican politics were reluctant to say the party's supporters had an aversion to "too many facts"—and the pollster's guidance was incorporated into GOP candidates' campaign strategies. In fact, as the 2012 elections approached, the cycle was dominated by Republican voices who couldn't get enough of bumper sticker messaging.

Some of the competitive GOP presidential hopefuls struggled to even pretend to care about governing details. For example, in February 2012, for example, Rick Santorum delivered lengthy remarks on health care policy in which the former two-term senator insisted, with great sincerity, that in the Netherlands, elderly citizens are routinely "euthanized involuntarily at hospitals."

The argument was apparently intended as a policy critique of health care systems with socialized insurance—Santorum said the Dutch kill senior citizens for "budget purposes"—but the argument was hopelessly insane, and the Republican's campaign refused to even try to defend it.[13]

Meanwhile, Herman Cain, a former pizza company executive who was briefly a top-tier contender for the Republican Party's presidential nomination, embodied post-policy campaigning in ways that bordered on performance art. Cain didn't bother to deny his disregard for substance; he celebrated it.

In October 2011 Cain was asked what would happen if reporters asked if he knew who the president of Uzbekistan is. He replied, "When they ask me who's the president of Ubeki-beki-beki-beki-stan-stan, I'm going to say, you know, I don't know. Do you know?"[14] A month later, Cain declared with pride, "We need a leader, not a reader."[15]

This came on the heels of Cain's announcement that, if elected president, he would not "allow" Congress to pass any legislation longer than three pages.[16] His signature proposal—a "9-9-9" system featuring a 9 percent income tax, a 9 percent corporate tax, and a 9 percent sales tax—was widely panned by economists and tax experts as ridiculous. Cain not only said he didn't care, but also he vowed to make it nearly impossible for lawmakers to change the tax code after his plan passed—a promise the candidate felt comfortable making, since he knew so little about the American legislative process.[17]

Santorum and Cain were ultimately defeated in the nominating process by Mitt Romney, whose candidacy was burdened by post-policy problems of its own. The GOP nominee developed an unfortunate reputation for staggering dishonesty, which he earned through months of mendacious campaigning.[18] Many of Romney's ugliest lies—including the bogus and racially charged allegation that Barack Obama had "gutted the work requirement" in federal welfare laws—reflected an unmistakable ambivalence toward the substance of public policy. The more the Republican was told the claim was demonstrably false, the more Romney and his team made it the centerpiece of their 2012 pitch.

When the candidate made a more concerted effort to take governance seriously, it nearly always went badly. Romney spent months warning the public, "We keep on shrinking our Navy. Our Navy is now smaller than any time since 1917." It fell to the Democratic president to explain during a debate, "I think Governor Romney maybe hasn't spent enough time looking at how our military works. You mentioned the navy, for example, and that we have fewer ships than we did in 1916. Well, governor, we also have fewer horses and bayonets because the nature of our military's changed. We have these things called aircraft carriers where planes land on them. We have these ships that go underwater, nuclear submarines. And so the question is not a game of *Battleship,* where we're counting ships. It's what are our capabilities?"[19]

But for the most part, Romney spent much of the election cycle downplaying the importance of governing details altogether. When economists panned the effects of his tax plan, for example, the Republican defended it by saying his plan "can't be scored" because he'd hid so much relevant information.[20]

A *Washington Post* analysis added soon after, "A tax plan that can't be scored because it doesn't include sufficient details is not a plan. It's a gesture towards a plan, or a statement of intended direction, or perhaps an unusually wonky daydream. But it's not a plan."[21]

After criticisms of Romney's evasiveness grew louder, he explained in an interview with the *Weekly Standard,* "One of the things I found in a short campaign against Ted Kennedy was that when I said, for instance, that I wanted to eliminate the Department of Education, that was used to suggest I don't care about education. So I think it's important for me to point out that I anticipate that there will be departments and agencies that will either be eliminated or combined with other agencies. . . . So will there be some that get eliminated or combined? The answer is yes, but I'm not going to give you a list right now."[22] This was far less persuasive than he seemed to realize. The GOP candidate's argument, in effect, was that he *could* have explained his vision in more detail, but it might spark a public backlash, so he'd rely on secrecy as a way to win the election.

More than a few observers tired of the tactics. The *Boston Globe*'s Scot Lehigh wrote in June 2012 that Romney, as a winning gubernatorial candidate ten years earlier, "campaigned like the management consultant he had once been, digging deep into issues and proposing thoughtful plans based on his analysis of the facts." A decade later, Lehigh added, it was as if "an antimatter Mitt" was running for president, with vagueness that extended "to the heart of the Republican candidate's core proposals."[23]

Presidential campaigns in vibrant and healthy democracies are supposed to feature major-party candidates engaged in a spirited fight over how their country will be governed. Romney wanted no part of such a contest, unable to muster the interest or the courage.

The 2014 election cycle was a better year for GOP successes at the ballot box, but not for voters hoping to see the party take substance more seriously. Former Republican senator Scott Brown, running in New Hampshire just two years after failing to win a second term in Massachusetts, had an especially memorable interview with the Associated Press in early 2014 in which he said of his candidacy, "Do I have the best credentials? Probably not. Cause, you know, whatever."[24]

Several months later, during a televised debate, Brown was asked about the kind of steps he'd pursue to reduce carbon emissions. "I'm not going to talk about whether we're going to do something in the future," the Republican replied, apparently confused about the purpose of a political campaign.[25]

Brown wasn't the only member of his party lacking a coherent message ahead of the 2014 midterms. American job growth was nearing a decade-long high, Affordable Care Act enrollment was healthy, and congressional Republicans had no achievements of their own to run on. In June and July GOP insiders prepared to focus the bulk of their attention on conspiracy theories, including the deadly attack on the U.S. embassy in Benghazi, Libya, two years earlier.

As it turned out, that wasn't altogether necessary. A campaign issue, which shouldn't have been a campaign issue, arrived unexpectedly.

In August 2014 Republican congressman Phil Gingrey of Georgia, a former obstetrician gynecologist, helped get the ball rolling, insisting to the Centers for Disease Control and Prevention that he'd seen reports "of illegal immigrants carrying deadly diseases," including the Ebola virus, into the United States.[26] Soon after, Todd Rokita, a Republican representative from Indiana (and an attorney, not an M.D.) made similar public comments.[27]

As concerns over Ebola spread, and Election Day drew closer, Republicans' post-policy tendencies overwhelmed any sense of judgment or propriety. At times, GOP officials weren't even sure what their talking points were supposed to be: some Republicans were outraged that the Democratic White House wasn't doing more to address the spread of the virus, while others in the party criticized the Obama administration for mobilizing too robust of a response. For example, former Florida congressman Allen West dismissed Ebola as nothing more than a "really bad flu bug."[28]

Soon after, Senator Rand Paul spoke with Glenn Beck on the conservative host's radio show and suggested the public was insufficiently frightened. "I do think you have to be concerned," Paul said. "It's an incredibly transmissible disease that everyone is downplaying, saying it's hard to catch. . . . I'm very concerned about this. I think at the very least there needs to be a discussion about airline travel between the countries that have the raging disease."[29]

As Election Day neared, the Kentucky Republican's eagerness to exploit public anxieties started to spin out of control. Paul publicly questioned Ebola assessments from the actual experts, blamed "political correctness" for the Ebola threat, and traveled to battleground states questioning whether Obama administration officials had the "basic level of competence" necessary to maintain public safety.[30]

He added soon after, describing a hypothetical flight, "If this was a plane full of people who were symptomatic, you'd be at grave risk of getting Ebola. If a plane takes twelve hours, how do you know if people will become symptomatic or not?" The senator went on to say the public could be at grave risk if fellow travelers on airplanes are "vomiting all over you or they're coughing all over you."[31]

Though Paul was more unhinged than most, he had plenty of company within his party. Wisconsin senator Ron Johnson, before taking over as chairman of the Senate Homeland Security and Governmental Affairs Committee, raised the prospect in late October 2014 of ISIS terrorists deliberately infecting themselves with Ebola, traveling to the United States, and killing an untold number of Americans.[32] "I think that is a real and present danger," Johnson said in an interview with a conservative media outlet.[33]

South Carolina congressman Joe Wilson told the public that members of the Palestinian militant group known as Hamas might transport Ebola into the United States through Mexico.[34] Scott Brown opined on Fox News that Americans "would not be worrying about Ebola right now" if Mitt Romney were in the Oval Office.[35] And Long Island's Peter King said in an interview on a local radio station that medical professionals weren't necessarily trustworthy on Ebola-related information because "maybe this is a mutated form of the virus."[36]

Republican Joni Ernst, just a few days before winning her U.S. Senate campaign in Iowa, complained that Barack Obama was "just standing back and letting things happen. . . . With Ebola, he's been very hands off." Ernst added that, as far as she was concerned, the Democratic president hadn't "demonstrated" whether he cared about Americans' safety.[37]

Private citizen Donald Trump, true to form, published tweets urging the public to blame the president for the threat,[38] going so far as to call for Obama's resignation.[39]

This came on the heels of one GOP congressman, Georgia's Paul Broun, urging conservatives concerned about Ebola to send him campaign contributions.[40]

On Capitol Hill, fellow Floridians Representative Dennis Ross and Senator Marco Rubio announced plans to introduce legislation in their respective chambers to prohibit travel between the United States and countries in western African.[41]

The party's hysterical response was a case study in post-policy politics at its most craven. At times, it was far from clear whether Republicans genuinely believed their own fearmongering, or whether they simply hoped public panic might give them an electoral edge on the eve of an important and consequential midterm cycle. Either answer was equally unsatisfying: GOP officials ignored every guidance from public health officials and took every opportunity to heighten public anxieties, peddling warnings that bordered on deranged.

Five days before Election Day 2014, Barack Obama tried to calm frayed nerves, delivering unscripted remarks from the White House.

"A lot of people talk about American exceptionalism. I'm a firm believer in American exceptionalism. You know why I am? It's because of folks like this," the Democrat said, pointing to a group of assembled medical professionals who'd helped combat Ebola. "It's because we don't run and hide when there's a problem. Because we don't react to our fears, but instead, we respond with common sense and skill and courage. That's the best of our history—not fear, not hysteria, not misinformation. We react clearly and firmly, even when others are losing their heads. That's part of the reason why we're effective. That's part of the reason why people look to us."[42]

Immediately after the elections, as Republicans celebrated having taken control of the U.S. Senate and growing their U.S. House majority, the GOP promptly lost interest in Ebola. The need to exploit public anxiety had evaporated: the number of American patients with Ebola had returned to zero, and the Obama administration's response had been proven a success.

The Republicans whose manic reactions to the threat were

exposed as ridiculous never explained why they'd failed the governing test so spectacularly.

Less than a year later, the race for the GOP's 2016 presidential nomination got under way. Texas governor Rick Perry helped set an early tone for the contest, arguing, "Running for the presidency's not an IQ test."[43]

In the months that followed, the Texan's comments proved prescient, as Republican White House hopefuls repeatedly demonstrated an unsettling degree of apathy toward public policy and how government works. In March 2015, for example, with much of the party's base focused on immigration, former Florida governor Jeb Bush was asked whether he'd end the DACA protections created by Barack Obama. Bush struggled to answer—because he appeared to have no idea what DACA was, despite his purported interest in immigration policy.[44]

In August 2015, Carly Fiorina, seeking the presidency following a six-year stint as the CEO of Hewlett-Packard, fielded a question from a veteran who said he was having trouble getting a doctor's appointment through Veterans Affairs. The California Republican, five years removed from a failed U.S. Senate campaign, responded that she'd address concerns surrounding the VA by asking veterans to come up with a plan for her, at which point she'd invite the public to vote via phone on whether to implement the plan. "Press one for yes on your smartphone, two for no," she said, failing to elaborate what would happen if the policy proved unpopular.[45]

A month later, former Arkansas governor Mike Huckabee, unfamiliar with American Civics 101, insisted that the U.S. Supreme Court's infamous *Dred Scott v. Sandford* decision of 1857—a historically reviled ruling that upheld the right to own slaves, which was superseded by the Thirteenth and Fourteenth Amendments to the U.S. Constitution—"remains to this day the law of the land."[46]

Marco Rubio said in May 2015 that his counterterrorism policy would be guided by a movie catchphrase. "Have you seen the movie *Taken*?" the senator asked a group of voters. "Liam Neeson. He had a line, and this is what our strategy should be: 'We will look for you, we will find you, and we will kill you.'"[47]

How would that be different from the existing policy? Where would Rubio look that the United States wasn't already looking? He never expressed much of an interest in the details; the catchphrase was supposed to suffice. Writing for *Esquire* magazine, Charles Pierce described the "Rubio Doctrine" as "three banalities strung together in such a way as to sound profound and to say nothing."[48]

In August 2015 Wisconsin governor Scott Walker told Fox News viewers they should cut through the noise of the campaign, visit his official website, and "see the details" of his platform.[49] What the governor apparently didn't realize was that his campaign's online presence did not have an issues page, and his site featured no details on any of his policy positions.

Walker was hardly alone: Politico found that of the seventeen Republicans running for president, nearly half had no policy information published to their campaign websites.[50] Most of the others had issues pages, but they were featured as "afterthoughts, light on significant detail."

And then, of course, there was the GOP candidate who ended up dominating the party's nominating process.

As post-policy politics overwhelmed the Republican Party, most GOP officials demonstrated indifference toward the substance of governing, but Donald Trump pushed this dynamic to new depths, demonstrating a flamboyant laziness toward evidence, reason, and basic information on how government works.

As the first major-party presidential nominee in American history to have literally no experience in public service of any kind, Trump faced an unusual challenge in 2016: demonstrating to the electorate that he was prepared to govern, despite never having done it. The Republican instead went out of his way to do the opposite.

In November 2015, for example, enjoying his front-runner status, Trump talked to Bloomberg News about how impressed he was with his role restoring an ice-skating rink in New York City's Central Park—a symbol, he said, of what he could achieve as the nation's president. Asked for a specific example of the kinds of problems he could solve, the Republican said, "I'll give you one example: wars. Wars aren't getting done. It's the same thing."[51]

The interviewer noted that his critics might balk at the idea of comparing an ice-skating rink to resolving international military conflicts. Trump, unfazed, insisted, "It's all the same."

As the campaign progressed, the future president's familiarity with his own country's system of government did not improve. Asked in late March 2016 about his plans to fill a U.S. Supreme Court vacancy, Trump said he intended to look for a nominee who would "look very seriously" at Hillary Clinton's email server protocols, reinforcing fears that he not only intended to use the levers of federal power against his perceived political enemies, but also that he had no idea what the judicial branch of the American government does.[52]

Shortly after Trump secured his party's nomination in early May, a confidant to the candidate acknowledged that the celebrity-turned-politician was reluctant to flesh out a policy agenda. "He doesn't want to waste time on policy and thinks it would make him less effective on the stump," the ally told Politico, adding that the GOP nominee intended to "figure out" his governing plans after Election Day, not before.[53]

This turned the model of how politics is supposed to work in a functioning democracy on its head. Candidates are supposed to craft a governing vision before an election, not only to help battle test an agenda during a lengthy public process, and not only to tell Americans what they'd be voting for, but also to empower victors to claim a popular mandate once voters had their say.

The Republican nominee, however, didn't see the point. Indeed, Trump's contempt for the very idea of presidential campaigns working on policy ideas became one of the staples of his candidacy. In June 2016 the Republican called the Democratic campaign's focus on substance "crazy," adding that writing policy ideas is "a waste of paper." Trump went on to tell *Time* magazine, "My voters don't care, and the public doesn't care."[54]

When the GOP nominee's team tried to prepare documents to help get him up to speed on issues he knew nothing about, Trump admitted, without embarrassment, that he saw no need to read them. Trump boasted to the *Washington Post* in July that he rarely reads at all because, as far as the then-candidate was concerned, he knew how to reach the right decisions "with very little knowledge other than the knowledge I [already] had."[55]

The article added, "Trump said reading long documents is a waste of time because he absorbs the gist of an issue very quickly. 'I'm a very efficient guy,' he said. 'Now, I could also do it verbally, which is fine. I'd always rather have—I want it short. There's no reason to do hundreds of pages, because I know exactly what it is.'"

Allan Lichtman, a political historian at American University, noted, "We've had presidents who have reveled in their lack of erudition. But Trump is really something of an outlier with this idea that knowing things is almost a distraction."[56]

There is a memorable scene in *The Simpsons Movie* from 2007 in which President Schwarzenegger is alerted to a crisis in Springfield. A White House aide presents him with five folders, each of which includes a different response. But the fictional president can't be bothered to review any of them.

"I was elected to lead, not to read," he says before choosing the middle folder without having opened it.

The scene was intended as satire. Trump inadvertently made it the basis for his approach to policy making.

As the star of the reality TV show *The Apprentice* was poised to officially claim his party's nomination in 2016, Politico's Michael

Grunwald explained, "The candidate's flagrant indifference to the details of public policy is particularly remarkable. . . . He has boasted that his main policy adviser is himself, and the advisers he does have say he doesn't read briefing papers. He has mocked Hillary Clinton for surrounding herself with 'eggheads' and churning out reams of wonky government reform proposals."[57]

At the GOP's nominating convention, Trump's attitudes permeated the party's thinking. Newt Gingrich, a prominent surrogate for the future president, sat down with CNN's Alisyn Camerota, who pressed the former House Speaker on some of the false claims his party's nominee was peddling on a daily basis.

The anchor noted, for example, that violent crime rates nationwide fell during the Obama era, despite Trump's claims to the contrary. "The average American, I will bet you this morning, does not think crime is down, does not think they are safer," Gingrich said. Camerota responded, "But it is. We are safer, and it is down." Gingrich, incredulous, said, "No, that's just your view."[58]

The two went back and forth for a while, with the CNN host pointing to actual data from the FBI, and the former GOP leader arguing that facts and reason are less relevant than false perceptions. "As a political candidate," Gingrich concluded, "I'll go with how people feel, and I'll let you go with the theoreticians."[59]

It was emblematic of an absurd school of thought: credible evidence is fine, as far as it goes, but verifiable crime statistics from the FBI are no match for what unprincipled politicians can get people to believe.

Gingrich's posture summarized Trump's post-policy strategy. Confronted with reality, the Republican response is to effectively declare, "I don't care."

The Republican nominee's team did include a policy shop, but the candidate ignored it too, and the office shuttered in the weeks following the party's national convention.[60] One former Trump policy adviser called the entire initiative "a complete disaster."

It was around this time that Sam Clovis, Trump's national

policy adviser, conceded that Team Trump had made a deliberate choice to treat governing details as an afterthought in the campaign. The American electorate, he added, would be "bored to tears" if the GOP campaign even tried to focus on substance.[61] Mind you, it was Clovis's job to help guide the Republican ticket's direction on substantive issues.

To drive home the point, CNN's Brian Stelter noted that both of the major-party presidential nominees had websites with positions on issues, but Hillary Clinton's was twelve times longer than her Republican rival's.[62]

MSNBC's Chris Hayes observed in September 2016 that the prospect of a Trump presidency was, for all intents and purposes, a "black box." The host added, "No one, probably not even Trump, knows what the hell it looks like."[63]

The fact that voters couldn't see into the black box was part of a breakdown in how an electoral system that values governance is supposed to work. Nevertheless, Republican candidates were rewarded up and down the ballot, leaving the party in control of all the levers of power for the first time in a decade. The expectation was that GOP policy makers would advance their agenda, tout their accomplishments ahead of the 2018 midterm elections, and await a reward from voters.

It didn't quite work out that way. Shortly before Christmas 2017, as the Republicans' regressive tax plan was poised to clear Congress, Senate Majority Leader Mitch McConnell looked ahead to the 2018 cycle with confidence. "If we can't sell this to the American people, we ought to go into another line of work," he said.[64]

In the year that followed, GOP officials abandoned boasts about their only major legislative success for a simple reason: polls found the American mainstream had no use for the tax package. According to a survey commissioned by the Republican National Committee, the public, by a two-to-one margin, saw the policy as one that favored "large corporations and rich Americans" over "middle-class families."[65] The report concluded that the party had "lost the messaging battle," and as a result, references to the tax breaks largely disappeared from GOP campaign advertising.[66]

Theoretically, Republicans could have shifted their attention to their other priorities for 2019 and 2020, but, by and large, the party's legislative wish list was thin and filled with wildly unpopular ideas that would leave the party in an even worse position if shared with the public. After GOP officials had waited a decade to assume power in Washington, DC, their governing cupboard was bare.

Three weeks before Election Day 2018, Ramesh Ponnuru, a conservative commentator, wrote a piece for Bloomberg News noting that Republicans not only failed to pursue any meaningful goals in the months leading up to the midterms, but also they appeared wholly unwilling to present ideas for the next Congress. Republicans, Ponnuru explained, "still have no agenda."[67]

The result was an awkward dynamic: The Republican Party, ostensibly a dominant governing party, couldn't talk about its recent past and hadn't bothered to craft enough of a policy agenda to focus on its near future.

Since remaining silent wasn't much of an option in an election season, GOP incumbents and candidates spent months trying to convince voters that the prospect of Nancy Pelosi ousting Paul Ryan as Speaker of the House was scary,[68] and nonwhite immigrants were scarier.[69]

With time running out, and his party's prospects fading, Donald Trump turned frantic, pointing to imaginary "riots" he claimed were under way in California,[70] insisting that Democrats wanted to "destroy" the Medicare system,[71] and falsely accusing Democrats of financing a nonexistent "caravan" of Central American and Middle Eastern migrants whom the president said were planning an "invasion" of the United States.[72]

"It doesn't matter if it's a hundred percent accurate," a senior Trump administration official told the Daily Beast in late October

2018, summarizing an unfortunate post-policy posture. "This is the play."[73]

When one of the president's supporters sent explosive devices to Democratic leaders, progressive voices, and journalists, Trump was outraged—not by the attempted terrorism, but by the fact that "bomb stuff," as he put it, was interfering with his election-season messaging.[74]

With two weeks remaining before Election Day, the president rolled out a brand-new, eleventh-hour idea: he and his party would approve "a very major tax cut for middle-income people" before the votes were cast. Congressional Republicans and White House officials, Trump added, were working "around the clock" on the new tax policy, which he said was badly needed, despite the tax cuts the GOP had approved a year earlier.[75]

Asked about a possible time frame for this provocative new gambit, the president said on October 20, "I would say sometime around the first of November, maybe a little before that."

No one had the foggiest idea what he was talking about. Lawmakers weren't even on Capitol Hill in late October, and White House officials conceded privately that they were "mystified" by their boss's rhetoric.[76] It was plainly obvious that Trump had made it all up, but he couldn't come right out and admit that, so the president proceeded to talk about his plan as if it were real.

"We're putting in a resolution sometime in the next week, or week and a half, two weeks," he told reporters a couple of days after first broaching the subject. Trump boasted to the *Wall Street Journal* that he and his team had even come up with a way to make his new tax plan "revenue neutral based on certain things."[77] Outside of the president's overactive imagination, those "things" did not exist. Neither did the new tax plan.

When a White House reporter asked if he was perhaps referring to some kind of executive order on tax policy he intended to sign, Trump replied, "No, no, no. I'm going through Congress."

A reporter noted moments later, "But Congress isn't in session."

The president's pitch was borderline incoherent—it was never clear what he thought a "resolution" was—but Trump clung to the lie, repeating it at a campaign rally in Texas, and expecting voters to believe the fantasy.

Trump didn't just describe a proposal that existed only in his mind, he did so in a way that required him to willfully disregard the details of tax policy and ignore how bills become laws in his own country.

The November 1 deadline came and went, as did Election Day, and the "very major" tax plan never materialized. Treasury Secretary Steven Mnuchin appeared at a forum a month later and was asked about the plan that only existed in the president's imagination.

"I'm not going to comment on whether it is a real thing or not a real thing," he replied.[78]

As the president's attention shifted in earnest to the 2020 cycle, his interest in constructive governing, which was already weak, evaporated. In April 2019 Peter Nicholas wrote for *The Atlantic* that Trump's rallies were getting longer and more aggressive, and the president's grievances were "ever more pronounced," but the Republican's appearances seemed "untethered to any sort of strategy to drive a policy agenda ahead of the 2020 presidential election."[79]

That, of course, was because Trump had no real policy agenda. He was certain he wanted to win a second term; he was less sure why.

A couple of months later, at his official reelection campaign kickoff in Orlando, Florida, Trump excoriated Hillary Clinton seven times in thirty minutes, but he failed to present any kind of substantive agenda for another four years in the White House.

Shortly after the Trump left the stage, Dan Bongino, a conservative commentator, appeared on Fox News and declared, in reference to the president's remarks, "He absolutely blistered Hillary Clinton."[80] There was some truth to that: Trump spent an unnerving amount of time expressing his preoccupation with the former Democratic secretary of state, implicating her in a lengthy list of scandals and making a spirited argument that Clinton was unworthy of public support.

The trouble, of course, was that Clinton wasn't running for anything. Election Day 2016 had come and gone nearly a thousand days earlier. By the time Trump launched his reelection campaign, Clinton had been gone from public office for nearly seven years.

Common sense suggested that the incumbent president would have put a positive spin on his scandal-plagued first term and presented at least some kind of forward-thinking agenda for a second term. With no obvious frontrunner for the Democratic presidential nomination, Trump's June 2019 remarks also offered the Republican an opportunity to set the parameters for the dominant issues and themes of the 2020 cycle.

But Trump, sounding a bit too much like a middle-aged man who couldn't stop reminiscing about the time he won the big game in high school, kept his focus on the 2016 race, Clinton, her email server protocols, and related far-right conspiracy theories.

Partway through his remarks, the president said of his reelection campaign staff, "They cost a fortune, and they never give me any ideas."[81] Given what the public heard from Trump, that was very easy to believe; the president's vision was sorely lacking in meaningful objectives. It was hard not to wonder, though, who bore responsibility: the highly paid advisers who failed to give the president new ideas, or the Republican incumbent who was interested only in old ones.

A month later, in July 2019, Trump had a phone meeting with Ukrainian president Volodymyr Zelensky, in which the new allied leader hoped to secure U.S. support in the face of Russian aggression. Trump was open to the possibility, but he told his counterpart in Kyiv, "I would like you to do us a favor, though."[82]

As the world soon learned, Trump, erasing the line between U.S. foreign policy and his campaign interests, expected Zelensky to come up with dirt on former Vice President Joe Biden, who declared his candidacy in the spring. It quickly became the basis for one of the most notorious extortion schemes in American history: the White House would consider delivering congressionally approved military assistance to our vulnerable ally, but Trump first expected Ukraine to deliver campaign fodder he could use against a Democratic candidate who was ahead in the polls.

As 2019 came to a close, the American president's brazen abuse led to his impeachment, but it also helped bolster the post-policy thesis: instead of approaching Ukraine with a carefully examined foreign policy, the Republican conflated governing and campaigning in ways that threatened to bring his presidency to a premature end.

Writing for Slate in November 2019, Lili Loofbourow reflected on Congress's impeachment hearings and the parade of administration witnesses who helped shine a light on the White House's misdeeds. She explained, "Their testimony has incidentally illustrated how Trump's incessant, unrelenting narcissism warps his ability to execute the duties of his office and the extent to which that dysfunction has spread, hobbling institutions that we need intact."[83]

As the details of the scandal came into focus, the president's GOP allies on Capitol Hill struggled to even try to marshal a substantive defense of Trump's abuses or address the case on the merits. In fact, the Capitol Hill newspaper *Roll Call* reported in November 2019 that many key House Republicans, invited to participate in the fact-finding investigation, "have simply not shown up" for hearings and depositions.[84]

It was post-policy politics at its most farcical.

In February 2020, Senate Republicans rejected an effort to bring witness testimony into the president's impeachment trial—creating the first such trial in American history without witness testimony—before voting to acquit Trump of both charges. (Senator Mitt Romney of Utah was the only GOP senator to vote "guilty" on one of the impeachment articles, enraging the White House.)

Several Republicans who participated in the trial acknowledged the president's guilt, though they voted with their party anyway. The GOP, indifferent to evidence and unmoved by principles, bore little resemblance to a governing party throughout the ordeal, leaving no doubt about Republicans' sole focus: the acquisition and maintenance of partisan power.

CHAPTER 11

Bridging the "Wonk Gap"

The Road Ahead

IN THE FIRST SEASON of NBC's *The West Wing*, there was an episode in which Sam Seaborn, a presidential speechwriter, has a lengthy argument with Mallory O'Brien, a schoolteacher and the daughter of the White House chief of staff. Their quarrel, spanning much of the hour, is over taxpayer-funded vouchers to be used to pay private school tuition. Sam, played by Rob Lowe, needles Mallory (Allison Smith) with pro-voucher arguments, leaving her offended and annoyed, especially after she sees his written position paper on the issue.

Eventually she takes her concerns to her father, Leo McGarry (John Spencer), who explains that Sam isn't an actual voucher advocate; he was simply engaged in an exercise called "opposition prep." The fictional chief of staff explains, "When we're gearing up for a debate, we have the smart guys take the other side."

At face value, this made perfect sense. Ahead of a policy fight, it stands to reason that political pugilists would prepare their strongest possible case, testing assumptions, anticipating likely criticisms, cultivating supporting evidence and background materials, and generally striving for a bulletproof presentation. All of this is built on the assumption that stronger policy arguments will beat weaker ones, and those in positions of authority, before settling on a position, will make sincere evaluations based on the merits.

In a televised drama, it made for engaging fictional scenes. In contemporary American politics, those scenes were fanciful to the point of comedy.

The Republican Party's transition into a post-policy party renders that entire model of governance almost meaningless. In practically every modern political dispute, GOP officials are wholly unconcerned with whose arguments are the most persuasive or which constituencies have the facts on their side. The party's principal concerns are electoral, political, and ideological, not practical.

For example, it's difficult for even the most naïve political observers to imagine Republicans, gearing up to pass tax breaks for the wealthy, questioning long-held assumptions and preparing compelling answers to Democratic objections. GOP lawmakers can't even bring themselves to hold congressional hearings to scrutinize the effects of their own proposals before voting on them. Asking "the smart guys" to "take the other side" in anticipation of a drawn-out policy debate is a step that never enters the equation. For those who follow current events closely, the idea is plainly absurd.

Democrats remain largely wedded to the traditional model. In November 2014 Barack Obama delivered a speech announcing new protections for immigrants, shielding them from deportation threats, and the president spent a significant chunk of his remarks debunking what his critics were likely to argue, explaining to the public in advance why they were wrong.[1]

Republicans, Obama predicted, would say the Democratic administration had been lax on border security—so the president reminded Americans that he'd increased border security, and illegal border crossings had fallen to a four-decade low. Republicans would say the president hadn't worked in a bipartisan way with Congress, so Obama reminded the public that he'd worked with lawmakers on the popular and bipartisan Gang of Eight bill—which House GOP lawmakers then refused to even consider. Republicans would say Obama's actions were unprecedented—so he explained how similar his actions were to steps taken by previous presidents from both parties.

And the Democratic president predicted that Republicans would say his policy constituted "amnesty," which led Obama to explain, "Amnesty is the immigration system we have today: millions of people who live here without paying their taxes or playing by the rules, while politicians use the issue to scare people and whip up votes at election time. That's the real amnesty—leaving this broken system the way it is."

After the speech, the White House's GOP detractors said Obama had been lax on border security, partisan and uncooperative, and was pursuing an unprecedented "amnesty" scheme. The president intended to present an air-tight case. To borrow the phrasing from *The West Wing*, officials had done their "opposition prep." It didn't matter.

Two years later, Donald Trump was elected president, and the Republican Party's indifference to the substance of governing became even more acute.

To be sure, the GOP president spent parts of his term giving the *appearance* of caring about affairs of state. The former reality-show personality was mindful of how his actions might play out on Americans' television screens, which led him to occasionally script some dramas for the cameras.

In June 2017, for example, Trump hosted a White House event featuring a fake signing ceremony in honor of the president's principal infrastructure goal: privatizing the nation's air-traffic control system.[2] The gathering had all the trappings of a major bill signing ceremony—Trump even surrounded himself with Republican members of Congress, who were only too pleased to accept ceremonial pens—except the president didn't sign any legislation or executive orders.

Rather, he put his signature on a glorified press release the White House labeled a "decision memo," which is a phrase Team Trump apparently made up. In practical terms, the president organized a theatrical event in which he signed a document asking lawmakers to replace air-traffic controllers with private-sector employees. He could have just as easily published a tweet with the same appeal.

Congress ignored the request.[3] The "decision memo" signing ceremony was theater with no purpose other than offering Trump an excuse to effectively tell the public, "Look everyone, I'm presidenting!"

The Republican's approach was similar when it came to his liberal use of executive orders. As a candidate, Trump told CNN in January 2016 that he was uncomfortable with the "executive order concept," adding, "You know, it's supposed to be negotiated. You're supposed to cajole, get people in a room, you have Republicans, Democrats, you're supposed to get together and pass a law. [Barack Obama] doesn't want to do that because it's too much work. So he doesn't want to work too hard. He wants to go back and play golf."[4]

In hindsight, the irony is staggering. Once in office, Trump barely tried to negotiate bipartisan deals, avoided too much work, played an inordinate amount of golf, and signed executive orders at the fastest clip of any president in thirty years.[5]

But what was amazing about most of the orders was that they existed only to give the *appearance* of action. In May 2019, the *Los Angeles Times* published an analysis that characterized the vast majority of Trump's executive orders as having little to no substantive value: "For a president who relishes pomp and shows of executive action, unchecked by Congress, signing ceremonies have become a hallmark, a way to convey accomplishment."[6]

While nearly all of the orders were signed with a degree of spectacle, the *Times* found that many of them simply directed cabinet secretaries to pursue a course of action, something Trump could easily have accomplished "with a phone call."[7]

Elaine Kamarck, the director of the Center for Effective Public Management at the Brookings Institution, a nonpartisan think tank in the nation's capital, added, "You don't really need an executive order for a lot of this stuff, but it makes for a good show. . . . He even gives out pens, which is really sort of ridiculous."

Foolish or not, putting on "a good show," as opposed to overseeing good governance, has become a staple of Trump's presidency. At times, that included the Republican playing the role of casting director.

In April 2019, for example, in remarks to the National Republican Congressional Committee, Trump praised NRCC chair Tom Emmer, a congressman from Minnesota, because he was "central casting." The president added, "You couldn't pick a better guy in Hollywood. There's no actor that could do it better."

In the same speech, Trump said he'd recently spoken to a U.S. general who was "out of central casting, I'm telling you. There's nobody in Hollywood that could look like this guy." A few days later, Trump traveled to Southern California, where he praised U.S. border officials as "central casting."

The fact that this was one his favorite phrases told the public something significant about how its commander in chief approached his duties. When the White House had to choose a Supreme Court nominee in 2018, Trump considered physical appearances important. "Beyond the qualifications, what really matters is, does this nominee fit a central-casting image for a Supreme Court nominee, as well as his or her spouse," a Republican close to the White House told Politico. "That's a big deal. Do they fit the role?"[8]

A few months earlier, Trump chose Ronny Jackson, his White House physician, to oversee the Department of Veterans Affairs, in part because of the Navy admiral's guise. "He's like central casting—like a Hollywood star," Trump told donors at a fund-raiser.[9] The president later conceded that Jackson "might not have been qualified" to lead the massive VA bureaucracy, responsible for the well-being of America's veterans, though Trump, unmoved by the relevant of qualifications, didn't much care.[10] (Jackson's nomination couldn't overcome bipartisan opposition, but in the wake of his failure, the doctor launched a 2020 congressional campaign in Texas.)

What mattered was the expectation in his mind about what an official was supposed to look like. When describing Vice President Mike Pence, Trump routinely said the Indiana Republican was "central casting."[11] On the day of his inauguration, the forty-fifth president pointed to Defense Secretary James Mattis and said, "This is central casting."[12] When Trump considered Mitt Romney for his cabinet, transition officials said the incoming president believed Romney "looks the part of a top diplomat right out of 'central casting.'"[13]

The *Washington Post*'s Karen Tumulty told MSNBC after the 2016 election that "central casting" is "actually a phrase [Trump] uses quite a bit behind the scenes."

This surprised no one. Trump cared about "central casting" as if he were the executive producer of an elaborate show—because to a certain extent, that's how the post-policy president perceived his role. Governing was an afterthought.

The passive indifference toward policy making was hardly limited to Trump. After the 2018 midterms, Republicans still controlled the U.S. Senate and the White House, but with Democrats enjoying a U.S. House majority, GOP leaders decided to effectively stop legislating ahead of the 2020 cycle.

As Senate Majority Leader Mitch McConnell brought the upper chamber to a crawl, and the Republican majority expressed little interest in tackling any of the challenges facing the nation, the Senate became known as a legislative "graveyard."[14] Democratic senator Chris Murphy complained that McConnell had "effectively turned the United States Senate into a very expensive lunch club that occasionally votes on a judge or two."[15]

To be sure, it wouldn't have been easy for a GOP-led Senate and a Democratic-led House to negotiate on any of the major issues of the day, but there was a self-proclaimed world-class deal maker in the Oval Office, and the nation had a lengthy list of challenges in need of attention. McConnell and his party's leadership team preferred not to try. The more House Speaker Nancy Pelosi shepherded through major proposals—in 2019, her chamber passed bills on issues ranging from voting rights to civil rights, climate change to gun safety—the more Senate Republicans collectively shrugged their shoulders and let the measures wither on the vine.

It would be a mistake to assume Republicans are incapable of effective policy making. When utilizing gerrymandering techniques to give GOP candidates unfair and undemocratic electoral advantages, for example, Republicans have demonstrated an uncanny ability to carefully utilize evidence in pursuit of substantive goals.

In North Carolina, for example, an appeals court found that GOP officials drew legislative district boundaries that targeted African-American voters "with almost surgical precision."[16] It was part of a racist scheme to undermine democracy, but it was also a reminder that when Republicans are willing to set their minds to it, they're able to roll up their sleeves and do substantive work in pursuit of their goals. Those goals are sometimes abhorrent, but the party at least has the capacity.

But outside of self-serving political schemes, the GOP has abandoned its status as a serious policy-making entity. The shift, however, need not be permanent. The Republican Party used to be a governing party, and it may yet become one again.

It could start by hiring some policy staff and, in the process, improve the party's capacity to govern. In 2014 Paul Glastris and

Haley Sweetland Edwards wrote an insightful cover story for *Washington Monthly* magazine, explaining in detail how Republican lawmakers had "made Congress stupid" by gutting Capitol Hill staff.[17] The deliberate shift began with the Gingrich Revolution in the mid-1990s, when the new GOP majority went on a firing binge, getting rid of lawyers, economists, investigators, auditors, analysts, and, perhaps most notably, subject-matter experts.

But GOP lawmakers reclaimed power after the 2010 elections, at which point the party cut even deeper, leaving Congress ill-equipped to deal with the complexities of policy making and governing in the twenty-first century. It's not that the halls of Capitol Hill grew empty, free of scurrying staffers. Rather, Republicans dismissed policy-making aides and hired communications staffers—whose principal responsibilities included writing talking points and raising their bosses' media profiles. GOP lawmakers, quite literally, decided to become less invested in governing and more invested in public relations.

USA Today reported that between 2011 and 2014, with Republicans in control of the U.S. House, policy-making staff shrunk by nearly 20 percent, while press and communications staff grew by nearly 15 percent. As the article summarized, "Members of Congress are putting your money where their mouths are."[18]

A year later, when Paul Ryan ascended to the Speaker's office, the GOP congressional leader didn't expand his policy staff, but he did hire a dozen taxpayer-funded Republican media and communications professionals.[19]

All of this was the opposite of what a governing party is supposed to prioritize. As a structural matter, members of Congress, many of whom are elected with limited policy literacy, have busy schedules and limited time to learn the granular details of complex policies on which they're expected to vote. They need staffs to help fill in the gaps.

As Vox's Matt Yglesias explained, "On most issues, most of the time, most members of Congress are more or less blindly following talking points that they got from somewhere else and that they don't really understand." They take cues from like-minded allies, "and their staff hastily assembles some stuff to say about it."[20]

The result is a governing dynamic in which lawmakers increasingly turn to pundits for political guidance and lobbyists—who are only too pleased to familiarize themselves with the substantive details that affect their clients—for legislative guidance. Hiring qualified policy staff wouldn't necessarily convince Republicans to *care* about the substance of governing, but it would be a step in a constructive direction.

The GOP could also reconsider its approach to intellectual infrastructure. It wasn't long ago that the right's think tanks were ascendant, churning out reliable scholarship in great volume for Republican officials and candidates. Institutions such as the Heritage Foundation have since fallen on hard times, following a trajectory similar to that of their partisan allies, shifting their focus from policy work to political activism.

Writing for *The Atlantic* in 2013, Molly Ball highlighted Heritage's transformation "from august policy shop to political hit squad" and the consequences for the intellectual underpinnings of the conservative movement and the Republican Party: "Without Heritage, the GOP's intellectual backbone is severely weakened, and the party's chance to retake its place as a substantive voice in American policy is in jeopardy."[21]

A month later, longtime Utah senator Orrin Hatch appeared on MSNBC and reminisced about the Heritage Foundation's former role as a dominant player in shaping conservative policy ideas. "There's a real question in the minds of many Republicans right now, and I'm not just speaking for myself: Is Heritage going to go so political that it really doesn't amount to anything anymore?"[22]

All of this has contributed to what many observers have described as the "wonk gap" between the parties. The *New York Times*'s Paul Krugman defined the phenomenon as the GOP's "near-complete lack of expertise on anything substantive. . . .

The dumbing down extends across the spectrum, from budget issues to national security to poll analysis."[23] A commitment from Republicans and their conservative allies to overhaul the right's intellectual infrastructure would create a foundation on which a governing party could build.

Republicans can and should also take a fresh look at the party's approach to empiricism. For many years, federal policy makers from both parties relied on independent data from official sources such as the Congressional Budget Office, the Congressional Research Service, and the Joint Committee on Taxation. As the GOP has embraced a post-policy philosophy, it has also grown reflexively skeptical, and at times overtly hostile, toward information that conflicts with the party's political assumptions.

In some instances, Republican antagonism toward empiricism can be amusing, such as Donald Trump's phone call to the director of the National Park Service the day after his presidential inauguration, ordering officials to produce new photographs of the event because he didn't believe the images of sparse crowds he'd seen on television.[24]

Most of the time, however, the phenomenon is more exasperating. In February 2019 Trump held a Rose Garden press conference to announce that he was redirecting taxpayer funds, in defiance of Congress, to build border fencing. At the event, CNN's Jim Acosta asked the president to respond to the data published by the Trump administration that showed undocumented immigrants, on average, commit fewer crimes than native-born Americans.

"You don't really believe that stat, do you?" Trump replied, dismissing the data from his own Department of Homeland Security. "Do you really believe that stat?"

The Republican then turned to a group of family members who'd lost loved ones purportedly in crimes and accidents committed by undocumented immigrants. "What do you think? Do you think I'm creating something?" Trump asked rhetorically.

"Ask these incredible women who lost their daughters and their sons, okay?"[25]

There's no question that those families had suffered horrible losses, but their painful experiences did not negate the statistical evidence. It would be comparable to someone reminding the president that modern air travel has fewer fatalities than any other form of transportation, only to have Trump respond by turning to people who'd lost loved ones in plane crashes. The people in this hypothetical deserve sympathy, of course, but their grief doesn't disprove what's plainly true about the safety of flights.

It was problematic to see Trump reject important and specific data compiled by his own team, but it was worse to see the president reject the very idea that statistical evidence carries greater weight than a series of tragic anecdotes.

At the same press conference, *Playboy* magazine's Brian Karem pressed Trump on his sources of information, since he was inclined to overlook the evidence compiled by the Department of Homeland Security. "I get my numbers from a lot of sources," the president responded vaguely, adding, "I use many stats. I use many stats." The Republican refused to elaborate with specifics.

There was no mystery as to why: Trump wanted to believe that immigrants commit more crimes than native-born Americans, so he decided his false belief was true. Pressed to support his assertions, and presented with evidence that proved his beliefs wrong, the president grew visibly agitated, apparently because he expected others to simply accept his falsehoods as fact, as he'd already done.

Indeed, in this case, Trump found it absolutely necessary to trust his gut over the data in order to prevent his house of cards from collapsing: if immigrants aren't responsible for rampant crime, then there was no crisis, and if there was no crisis, then there was no need for an executive order empowering the White House to circumvent Congress's appropriations process.

Ergo, Trump decided to leave empiricism to the data nerds

and belittle those who had the audacity to bother him with inconvenient truths.

An approach to governance that rejects the value of independent evidence will inevitably fail. Facts may be burdensome at times, and reality will occasionally prove disappointing, but removing empiricism from the policy-making process is folly.

"If there's no truth, how do we discuss and make decisions that are rooted in fact?" Rob Stutzman, a Republican operative and Trump critic, said in 2018, adding, "It is a huge fundamental problem of how to govern when there are no facts."[26]

The most effective impetus to changing the GOP's post-policy posture, however, will be external pressures that leave Republican leaders with the impression that the status quo is untenable. Some of this should come from independent news organizations that have too often been negligent, signaling to the party and voters that policy is trivia of limited importance to the American mainstream.

Much has been said about the media's coverage of the 2016 presidential election, but the *Columbia Journalism Review* published a report a year after Election Day that jolted many who saw it. *CJR* found that between October 29, 2016, and November 3, 2016—a six-day span that included early voting in much of the country—the *New York Times* published "as many cover stories about Hillary Clinton's emails as it did about all policy issues combined in the 69 days leading up to the election."[27]

This was evidence of a public discourse that wasn't nearly as healthy as it should have been. Just as importantly, though, this also isn't the sort of thing that happens in a political ecosystem that recognizes the proper role issues should play in national campaigns for the world's most powerful and most difficult job.

But throughout American history, the most powerful force for partisan change has been the incentive to win elections. Major parties overhaul their tactics and perspectives not after victories but in the wake of multiple defeats. With this in mind, the Republican Party is likely to become a governing party again when American voters tell the GOP it has no other choice.

Public tolerance for the party's policy lethargy has, for all intents and purposes, allowed Republicans to get away with their regressive transition. Many of the party's own voters have been content to overlook GOP officials' unconcern for governing, satisfied that the party's ideological predispositions are enough.

They're not. As the Republican Party drifts further from its governing roots, unable to effectively evaluate and execute its own plans, GOP voters aren't watching conservative technocrats implement conservative solutions. Rather, they see a party flailing, unwilling and unable to craft a clear vision, scrutinize their own ideas, and implement a policy agenda the way a competent governing party should.

Republican voters are hardly the only ones who should be alarmed. The American policy-making process, dependent on constructive contributions from major parties, has grown sclerotic and unresponsive. Every opportunity for constructive policy making follows a tired trajectory in which the parties talk past each other, with Democrats pushing their rivals to support their contentions with evidence and scholarship, and Republicans regurgitating talking points untethered to substance.

Even when GOP officials manage to pass major legislation, such as their regressive 2017 tax plan, the results are too often predictable failures. Promises go unmet due largely to ideas that went unexamined. The party's policy-making process—no hearings, no analyses, curtailed debate—is difficult to watch through uncovered eyes.

When House Speaker Paul Ryan conceded, in the wake of the demise of one of his party's health care endeavors, that his GOP was still struggling to become "a governing party," his candor was welcome, but his observation was unacceptable. At the time of Ryan's comment, the Republican Party had controlled the U.S. House for eighteen of the previous twenty-two years. The GOP had held a majority in the U.S. Senate for nearly as long. Over the last half century, nine men have sat in the Oval Office; only three of them were Democrats.

The party has not been without opportunities to learn how to govern responsibly in the world's dominant superpower; elected Republican officials simply haven't taken advantage of those opportunities, dismissing substantive due diligence as unnecessary and unsatisfying. If reliable data was going to tell GOP officials to pursue smarter policy options, the party preferred to go without.

Ours is an ailing political system that needs more than one governing party to recover. That will not and cannot happen until GOP officials recognize their post-policy shortcomings and take steps to correct them. The question is whether Republicans are prepared to change direction and value anew the importance of substantive policy making.

A great deal is riding on their answer.

ACKNOWLEDGMENTS

MY MOST SINCERE THANKS to Peter Hubbard, Molly Gendell, Alison Hinchcliffe, Kayleigh George, Stephanie Vallejo, and the amazing team at William Morrow and HarperCollins. This book would not be possible without them.

I also can't imagine what I would have done without Laurie Liss at Sterling Lord Literistic, who helped me navigate a complex and unfamiliar publishing process.

My gratitude is endless for Rachel Maddow—whom I consider both a friend and a hero—who has been generous to me in ways I will never be able to repay. I also want to extend my heartfelt thanks to the rest of my MSNBC family, including Cory Gnazzo, Laura Conaway, and Will Femia, for their extraordinary support.

I've also been fortunate to have an extraordinary group of friends, neighbors, and family members, who've been listening to me rant about my post-policy thesis for years. A special thanks to Michael Weitzner, Zoe Poulson, Rob Boston, Bill Simmon, Emily Stoneking, Alex Wolfson, Frederick Buckland, Bill Wolff, Greg Sargent, and E. J. Dionne for their encouragement.

Finally, and most importantly, I want to thank my wife, Eve, and my mom, Gini, who mean more to me than I could ever say.

NOTES

CHAPTER 1: "We're Not Great at the Whole Governing Thing": Meet the Post-Policy Party

1. Aiden Quigley, "Ryan Issues Namath-like 'Guarantee' on Obamacare Repeal," *Politico*, last modified March 7, 2017.
2. Stephanie Armour, Siobhan Hughes, and Kristina Peterson, "House GOP Leaders Surprised by Conservative Opposition to Health Plan," *Wall Street Journal*, March 9, 2017.
3. Alex Roarty (@Alex_Roarty), "GOP aide on CapHill: "I'm starting to think that while we're pretty good at winning elections, we're not great at the whole governing thing,'" Twitter, March 24, 2017, 1:31 P.M.
4. Neil Irwin, "Why the Trump Agenda Is Moving Slowly: The Republicans' Wonk Gap," *New York Times*, February 28, 2017.
5. John DiIulio, "John DiIulio's Letter," *Esquire*, last modified May 23, 2007.
6. Alex Isenstadt, "GOP Turns to Bush Aides for Advice," Politico, last modified May 4, 2009.
7. Matt Fuller, "Sessions Eyes Elections—and Beyond," *Roll Call*, last modified November 5, 2013.
8. Carl Hulse and Adam Nagourney, "Senate G.O.P. Leader Finds Weapon in Unity," *New York Times*, March 16, 2010.
9. Joshua Green, "Strict Obstructionist," *The Atlantic*, January/February 2011.
10. Alexis Simendinger, "In His Own Words: Mitch McConnell," *National Journal*, March 20, 2010.
11. Major Garrett, "Top GOP Priority: Make Obama a One-Term President," *National Journal*, October 23, 2010.

12. Michael O'Brien, "Boehner: Judge Congress by How Many Laws It Repeals, Not Passes," NBC News online, last modified July 21, 2013.
13. The White House Office of the Press Secretary, "Remarks by the President [Barack Obama] on the Economy—Northwestern University," news release, October 2, 2014.
14. Alex Shephard, "The Trump Campaign Will Never End," *New Republic*, December 1, 2016.
15. Steve Benen, "In Picking Cabinet, Trump Values Wealth, Inexperience," MSNBC online, last modified November 30, 2016.
16. Nancy Cook and Andrew Restuccia, "Trump Marginalizes D.C. Transition Staff," Politico, last modified December 6, 2016.
17. Matthew Lee, "Day Before Inauguration, State Department Lacks Interim Boss," Associated Press, last modified January 19, 2017.
18. Dan De Luce and John Hudson, "Trump's National Security Team Is Missing in Action," *Foreign Policy*, January 18, 2017.
19. Rich Lowry, "The Crisis of Trumpism," Politico, last modified March 29, 2017.
20. Brian Beutler, "Don't Lose Sight of the Trump Presidency's Real Problem," *New Republic*, last modified April 10, 2017.
21. Maggie Haberman and Glenn Thrush, "Trump Reaches Beyond West Wing for Counsel," *New York Times*, April 22, 2017.
22. Stephen J. Adler, Jeff Mason, and Steve Holland, "Exclusive: Trump Says He Thought Being President Would Be Easier Than His Old Life," Reuters, last modified April 27, 2017.
23. Philip Bump, "Donald Trump Pledges to Make Every Dream Possible, Which Seems Ambitious," *Washington Post*, September 28, 2016.
24. Ben Jacobs (@Bencjacobs), "Trump: 'we are going to fulfill every single wish and every single promise,'" Twitter, September 29, 2016, 3:31 P.M.
25. Aaron Blake, "President Trump's Full *Washington Post* Interview Transcript, Annotated," *Washington Post*, November 27, 2018.
26. Andrew Restuccia, Danial Lippman and Eliana Johnson, "'Get Scavino in Here': Trump's Twitter Guru Is the Ultimate Insider," Politico, last modified May 16, 2019.
27. Cristiano Lima, "Trump Boosts 'Space Force' Idea, Says U.S. Will Reach Mars 'Very Soon,'" Politico, last modified March 13, 2018.
28. Christina Wilkie (@christinawilkie), "If the Trump campaign's design for

a Space Force logo looks familiar, that's because it is.," Twitter, August 9, 2018, 4:17 P.M.

29. Tariq Malik, "Trump Unveils New Space Force Logo (Yes, It Looks Like Something From 'Star Trek')," Space.com, January 25, 2020.
30. Donald J. Trump (@RealDonaldTrump), "Space Force all the way!," August 9, 2018, 12:03 P.M.
31. David A. Graham, "Why the Space Force Is Just Like Trump University," *The Atlantic*, last modified August 10, 2018.

CHAPTER 2: "Manipulate the Numbers and Game the System": Economic Policy

1. Josh Marshall, "Painfully Stupid," Talking Points Memo, last modified, February 3, 2009.
2. Michelle Levi, "McConnell Says President Should Get Involved," CBS News online, last modified February 1, 2009.
3. "GOP Wants Spending Freeze," Daily Beast, last modified February 25, 2009.
4. David Weigel, "GOP Turns to Talk of Spending Freeze," Washington Independent, last modified February 24, 2009.
5. Ryan Powers, "Brooks: Boehner's Spending Freeze Would Be 'Insane,' GOP Is 'Stuck with Reagan,'" ThinkProgress, last modified March 8, 2009.
6. "Republicans Propose 'No Cost' Stimulus," Fox News online, last modified January 27, 2017.
7. Ben Armbruster, "Krugman: How Can There Be Bipartisanship When GOP 'Take Their Marching Orders from Rush Limbaugh?,'" ThinkProgress, last modified February 9, 2009.
8. Steven Pearlstein, "Wanted: Personal Economic Trainers. Apply at Capitol," *Washington Post*, February 6, 2009.
9. Josh Marshall, "A Turning Tide?," Talking Points Memo, last modified February 5, 2009.
10. Bernie Becker, "Sessions, Stimulus and the Taliban," *New York Times*, February 5, 2009.
11. Ezra Klein, "They Think We're Dumb. Otherwise, They Wouldn't Talk to Us Like This," *American Prospect*, last modified February 12, 2009.
12. Lindsey Graham, interview by Wolf Blitzer, *The Situation Room*, CNN, February 11, 2009.

13. Dan Carden, "Pence Sweetens on Stimulus as State Benefits," *Northwest Indiana Times* (Munster, IN), June 1, 2014.
14. *Estimated Impact of the American Recovery and Reinvestment Act on Employment and Economic Output from January 2011 Through March 2011* (Washington, DC: Congressional Budget Office, May 25, 2011).
15. Pat Garofalo, "Flashback: Republicans Warned That GM Rescue Was 'Road Toward Socialism,' 'Predictable' Disaster," ThinkProgress, last modified November 18, 2010.
16. Jonathan Weisman, "U.S. Declares Bank and Auto Bailouts Over, and Profitable," *New York Times*, December 19, 2014
17. David Shepardson, "Rubio Criticizes $85 Billion Auto Bailout," *Detroit News*, April 17, 2015.
18. Brian Beutler, "Boehner: If Jobs Are Lost as a Result of GOP Spending Cuts 'So Be It,'" Talking Points Memo, last modified February 15, 2011.
19. Ruth Marcus, "Boehner's Unreality Check on the Deficit," *Washington Post*, May 10, 2011.
20. Erik Wasson, "CBO: Obama Jobs Bill Reduces Budget Deficit," *The Hill*, October 7, 2011.
21. Associated Press, "Economists Show Support for Obama Job-Growth Plan," last modified September 10, 2011.
22. Erik Wasson, "Senate Republicans Present Obama with Counteroffer on Jobs," *The Hill*, October 13, 2011.
23. Greg Sargent, "Economist: Senate GOP Jobs Plan Wouldn't Help Economy in Short Term, and Could Even Hurt," *Washington Post*, October 14, 2011.
24. Eric Cantor, "An America That Works," *National Review*, last modified February 13, 2014.
25. Jay Bookman, "A 'Nothingburger' of a Policy Agenda from Eric Cantor," *Atlanta Journal-Constitution*, February 13, 2014.
26. Jonathan Chait, "Why Republicans Don't Want to Extend the Payroll Tax Cut," *New Republic*, last modified August 8, 2011.
27. Siobhan Hughes, "Ryan Downplays Prospects of New Deficit Committee," *Wall Street Journal*, August 7, 2011.
28. Brian Beutler, "Is the GOP Sabotaging the Economy? Schumer: 'They Give Us No Choice but to Answer Yes,'" Talking Points Memo, last modified July 1, 2011.
29. Tamara Keith, "Senate Votes to Keep Jobs Bill from Being Debated," National Public Radio, last modified October 12, 2011.

30. Alexander Bolton, "GOP Senators Angry About Reid Claim That They're Rooting for Economic Failure," *The Hill*, October 19, 2011.
31. Matt O'Brien, "Republicans Couldn't Possibly Be More Hypocritical About the Economy," *Washington Post*, May 15, 2019.
32. David Barstow, Susanne Craig, and Russ Buettner, "Trump Engaged in Suspect Tax Schemes as He Reaped Riches from His Father," *New York Times*, October 2, 2018.
33. Russ Buettner and Susanne Craig, "Decade in the Red: Trump Tax Figures Show over $1 Billion in Business Losses," *New York Times*, May 8, 2019.
34. "Transcript: Interview with Donald Trump," *Economist*, last modified May 11, 2017.
35. Merriam-Webster (@MerriamWebster), "The phrase 'priming the pump' dates to the early 19th century.," Twitter, May 11, 2017, 8:03 A.M.
36. Aaron Rupar (@atrupar), "Former Trump economic adviser Gary Cohn tells MSNBC, "I'm not an advocate of tariffs, & I'm not an advocate of trade wars." Suggests those were reasons he left the White House. Pressed by @SRuhle if Trump understands "basic economics," Cohn repeatedly dodges the question.," Twitter, January 24, 2019, 10:36 A.M.
37. Jeff Cox, "Former Fed Chair Janet Yellen Says Trump Has a 'Lack of Understanding' of Fed Policies and the Economy," CNBC online, last modified February 25, 2019.
38. Randall Lane, "Inside Trump's Head: An Exclusive Interview With the President, And The Single Theory That Explains Everything," *Forbes*, November 14, 2017.
39. Robert Farley, "Trump Inflates GDP Growth," FactCheck.org, last modified July 13, 2018.
40. Patricia Cohen, "U.S. Economy Grew at 2.6% Rate in Fourth Quarter," *New York Times*, January 26, 2018.
41. John W. Schoen, "Trump Defies Data with 6% GDP Growth Forecast," CNBC online, last modified December 6, 2017.
42. Steve Benen, "Job Growth Cooled a Little as 2019 Came to a Close," MSNBC online, January 10, 2020.
43. Brian Faler, "Camp Wages Own Battle Against ACA," Politico, last modified November 13, 2013.
44. "Top Frustrations with Tax System: Sense That Corporations, Wealthy Don't Pay Fair Share," Pew Research Center, last modified, April 14, 2017.

45. Mark Murray, "NBC/WSJ Poll: Public Likes Trump's Bipartisan Move—but Little Else," NBC News online, last modified September 21, 2017.
46. Christine Filer, "Two-Thirds Say Large Corporations Pay Too Little in Federal Taxes (Poll)," ABC News online, last modified September 26, 2017.
47. Steve Forbes, Larry Kudlow, Arthur B. Laffer, and Stephen Moore, "Why Are Republicans Making Tax Reform So Hard?," *New York Times*, April 19, 2017.
48. Shane Goldmacher, "How Trump Gets His Fake News," Politico, last modified May 15, 2017.
49. John Cassidy, "The Passage of the Senate Republican Tax Bill Was a Travesty," *New Yorker*, December 2, 2017.
50. Gail Collins, "Trump, Taxes, and . . . You Know," *New York Times*, November 1, 2017.
51. John Harwood, "Rep. Tom Cole doesn't trust the economists on GOP tax plan: 'There are about as many economists as there are opinions,'" CNBC, December 17, 2017.
52. Kailani Koenig, "Mulvaney: 'Gimmick' will help GOP pass tax reform," NBC News, November 19, 2017.
53. Lois Beckett, "White House would ditch attack on Obamacare in order to pass tax bill," *Guardian*, November 19, 2017.
54. Ruth Marcus, "The End of Shame," *Washington Post*, November 24, 2017.
55. Senator Chris Murphy (@ChrisMurphyCT), "Uhhh . . . we're hours away from a multi trillion dollar rewrite of the tax code and the people voting for it don't know what's in it.," Twitter, November 30, 2017, 1:21 P.M.
56. Senator Jon Tester (@SenatorTester), "I was just handed a 479-page tax bill a few hours before the vote. One page literally has hand scribbled policy changes on it that can't be read. This is Washington, D.C. at its worst. Montanans deserve so much better," Twitter, December 1, 2017, 7:06 P.M.
57. Josh Barro, "Something Very Stupid Is Happening in the Senate Right Now," Business Insider, last modified November 30, 2017.
58. White House, "Remarks by President Donald Trump, Vice President Mike Pence, and Members of Congress at Bill Passage Event," news release, December 20, 2017.
59. Derek Thompson, "The GOP Tax Cuts Didn't Work," *The Atlantic*, last modified October 31, 2019.
60. Reuters, "$1.5 Trillion Tax Cut Had No Major Impact on Business Spending," last modified January 28, 2019.

61. Heather Long, "GOP Leader Concedes Tax Cuts May Not Pay for Themselves as 2019 Deficit Grows," *Washington Post*, June 11, 2019.
62. Stephen Ohlemacher and Marcy Gordon, "Senate Passes GOP Tax Bill, Setting Stage for Final House Vote on Wednesday," Associated Press, last modified December 20, 2017.
63. Jim Tankersley and Karl Russell, "No One's Talking About the New Tax Law," *New York Times*, April 16, 2018.
64. Joseph Bankman, Daniel Hemel, and Dennis Ventry, "Why Filing Taxes Isn't Easy," Policito, last modified July 18, 2018.
65. Jordan Weissmann, "Senate Republicans Made a $289 Billion Mistake in the Handwritten Tax Bill They Passed at 2 A.M. Go Figure," Slate, last modified December 6, 2017.
66. Brian Faler, "Republican Tax Law Hits Churches," Politico, last modified June 26, 2018.
67. Eric Levitz, "Trump Accidentally Raised Taxes on the Children of Dead Veterans," *New York*, May 17, 2019.
68. Erica L. Green, "Low-Income College Students Are Being Taxed Like Trust-Fund Babies," *New York Times*, May 17, 2019.
69. Brian Faler, "'This Is Not Normal': Glitches Mar New Tax Law," Politico, last modified February 24, 2018.
70. "Tax Reforms," Initiative on Global Markets Questionnaire, May 2, 2017.
71. Tara Palmeri, "'The Cut Cut Cut Act': Trump, Hill Leaders Differ on Tax Overhaul Bill's Name," ABC News online, last modified November 2, 2017.
72. Josh Dawsey, "'Twenty Is a Pretty Number.' In Tax Debate, Trump Played the Role of Marketer in Chief," *Washington Post*, December 20, 2017.
73. Gerard Baker, Peter Nicholas, and Michael C. Bender, "Trump Eyes Tax-Code Overhaul, with Emphasis on Middle-Class Break," *Wall Street Journal*, July 25, 2017.
74. "Trump's Trade War," *Frontline*, written and directed by Rick Young, aired June 7, 2019, on Public Broadcasting Service.
75. Toluse Olorunnipa, "Kudlow Says TPP Reconsideration Is More of a 'Thought Than a Policy,'" Bloomberg, last modified April 17, 2018.
76. James Dean, Bruno Waterfield, and Oliver Wright, "Trump Puts EU Ahead of Britain in Trade Queue," *Times of London* (UK) online, last modified April 22, 2017.
77. Josh Dawsey, Damian Paletta, and Erica Werner, "In Fund-Raising Speech,

Trump Says He Made Up Trade Claim in Meeting with Justin Trudeau," *Washington Post*, March 15, 2018.

78. Jeet Heer, "Meet the New NAFTA, Same as the Old NAFTA," *New Republic*, October 1, 2018.
79. Paul Krugman (@paulkrugman), "My original prediction on Trump/NAFTA was that we would end up making some minor changes to the agreement, Trump would declare victory, and we'd move on. That's what seems to have happened 1/," Twitter, October 1, 2018, 7:43 A.M.
80. Michael C. Bender, "Trump Plans to Rebrand Nafta, Warns Canada," *Wall Street Journal*, September 13, 2018.
81. Justin Wolfers (@JustinWolfers), "A key non-negotiable U.S. requirement in the NAFTA renegotiations was that the revised agreement not be called NAFTA.," Twitter, October 1, 2018, 7:41 A.M.
82. Catherine Rampell, "Trump's 'Historic' Trade Deal Doesn't Look So Historic After All," *Washington Post*, October 1, 2018.
83. Stephanie Ruhle, "Trump Was Angry and 'Unglued' When He Started a Trade War, Officials Say," NBC News online, last modified March 2, 2018.
84. Ruhle, "Trump Was Angry."
85. White House, "Remarks by President Trump and Prime Minister Netanyahu of Israel Before Bilateral Meeting," news release, March 5, 2018.
86. Matt O'Brien, "Trump's Trade War Is an Intellectual Disaster," *Washington Post*, May 9, 2019.
87. Jim Tankersley, "Trump Hates the Trade Deficit. Most Economists Don't," *New York Times*, March 5, 2018.
88. Neal Rothschild, "Despite Promises, Trump's Trade Deficits Are Only Growing," Axios, last modified September 3, 2019.
89. Zeeshan Aleem, "Trump's Trade Tweets Prove One Thing: He Doesn't Understand Trade," Vox, last modified March 2, 2018.
90. Peter Coy, "After Defeating Cohn, Trump's Trade Warrior Is on the Rise Again," Bloomberg, last modified March 8, 2018.
91. Sarah Ellison, "The Inside Story of the Kushner-Bannon Civil War," *Vanity Fair*, last modified April 14, 2017.
92. Michael Hiltzik, "Column: Trump's Trade War Won't Achieve What He Wants, But It Will Hurt American Workers and Consumers," *Los Angeles Times*, April 5, 2018.
93. Victoria Guida, "Chinese Leaders 'Absolutely Confused' by Trump's Demands on Trade," Politico, last modified June 23, 2018.

94. Paul Krugman, "How to Lose a Trade War," *New York Times*, July 7, 2018.
95. Annie Lowrey, "The 'Madman' Behind Trump's Trade Theory," *The Atlantic*, December 2018.
96. Jonathan Swan, "Trump's Long Trade War," Axios, last modified May 14, 2019.
97. Swan, "Trump's Long Trade War."
98. Aaron Rupar (@atrupar), "'what..... WHAT!,'" Twitter, July 24, 2018, 1:07 P.M.
99. White House, "Remarks by President Trump in Thanksgiving Teleconference with Members of the Military," news release, November 23, 2018.
100. Matt O'Brien, "On Trade, Trump Either Doesn't Understand the Basic Facts or He Doesn't Care," *Washington Post*, April 18, 2018.

CHAPTER 3: "Even If It Worked, I Would Oppose It": Health Care

1. Karen Tumulty, "The Health-Care Talks: Will Obama Get More Involved?," *Time* online, last modified May 5, 2009.
2. Jonathan Cohn, "Republicans Would Repeal Obamacare in Precisely the Way They Accuse Democrats of Enacting It," HuffPost, last modified January 8, 2017.
3. Ezra Klein, "Do Democrats Have Their Own Individual Mandate?," *Washington Post*, June 18, 2012.
4. Fred Barbash, "Hatch: Mandate Is 'Totalitarianism,'" Politico, last modified March 25, 2010.
5. Igor Volsky, "Grassley: I Supported the Individual Mandate Before I Realized It Was Unconstitutional," ThinkProgress, last modified March 25, 2010.
6. Michael Scherer, "More Bad Republican Form," *Time* online, last modified September 21, 2009.
7. Carl Hulse, "In Lawmaker's Outburst, a Rare Breach of Protocol, *New York Times*, September 9, 2009.
8. Ezra Klein, "Sarah Palin's Ghostwriter Gets Hold of John McCain's Web Site," *Washington Post*, November 22, 2009.
9. Jodi Jacobson, "GOP Misinformation Machine Goes into High Gear on Senate Language on Abortion," Rewire.News, last modified November 19, 2009.
10. Kate Nocera, "Paul: 'Right to Health Care' Is Slavery," Politico, last modified May 11, 2011.

11. David Kurtz, "Harvard Ain't What It Used to Be," Talking Points Memo, last modified August 20, 2009.
12. Ruth Marcus, "Ruth Marcus on Debunking Health Reform Myths," *Washington Post*, November 11, 2009.
13. Robert Pear and David M. Herszenhorn, "Democrats Push Health Care Plan While Issuing Assurances on Medicare," *New York Times*, July 28, 2009.
14. Jason Linkins, "Virginia Foxx Warns That Health Care Reform Will Involve Widespread Elder Murder," HuffPost, last modified August 28, 2009.
15. Ezra Klein, "Is the Government Going to Euthanize Your Grandmother? An Interview with Sen. Johnny Isakson," *Washington Post*, August 10, 2009.
16. Sam Stein, "Grassley Endorses "Death Panel" Rumor: 'You Have Every Right to Fear,'" HuffPost, last modified September 12, 2009.
17. Amanda Terkel, "Grassley Blames Obama for Making Him Say That Health Care Reform Would 'Pull the Plug on Grandma,'" ThinkProgress, last modified August 23, 2009.
18. Steven T. Dennis, "Republicans Unveil Health Plan but Are Thin on Details," *Roll Call*, last modified June 17, 2009.
19. Alex Isenstadt, "Republicans Outline Health Plan," Politico, last modified June 17, 2009.
20. Molly K. Hooper, "Republicans impatient with leaders, awaiting healthcare alternative," *The Hill*, October 27, 2009.
21. Jonathan Chait, "Why You Can't Discuss Health Care with the GOP," *New Republic*, last modified February 25, 2010.
22. Matthew Yglesias, "Boehnercare: No Soup for You," ThinkProgress, last modified November 3, 2009.
23. Ben Armbruster, "CBO Says GOP Health Care 'Alternative' Leaves 52 Million Uninsured by 2019," ThinkProgress, last modified November 5, 2009.
24. Greg Sargent, "The Morning Plum: Obamacare May Perish. What Would GOP Do About Tens of Millions of Uninsured?," *Washington Post*, April 4, 2012.
25. Sahil Kapur, "Republicans Warn of New Obamacare Reality: No Repeal Without Alternative," Talking Points Memo, last modified April 8, 2014.
26. Chris Riotta, "GOP Aims to Kill Obamacare Yet Again After Failing 70 Times," *Newsweek*, last modified July 29, 2017.

27. Dana Milbank, "On Obamacare, Republicans Test the Definition of Insanity," *Washington Post*, July 17, 2013.
28. Jeremy W. Peters, "House to Vote Yet Again on Repealing Health Care Law," *New York Times*, May 14, 2013.
29. Manu Raju, "A New Dem Threat to Health Care Law," Politico, last modified February 7, 2011.
30. Raju, "A New Dem Threat."
31. Kate Nocera and J. Lester Feder, "CBO: Health Law to Shrink Workforce," Politico, last modified February 10, 2011.
32. Glenn Kessler, "Playing Games with CBO Testimony on Jobs and the Health-Care Law," *Washington Post*, February 11, 2011.
33. Steve Benen, "Even the Right Can't Deny Job Market's Hot Streak," MSNBC online, last modified January 8, 2016.
34. Amanda Terkel, "Sen. John Barrasso: White House Is 'Cooking the Books' on Obamacare Enrollment Numbers," HuffPost, last modified March 30, 2014.
35. Office of the Speaker of the House, "Boehner Statement on President's Health Care Law," news release, April 1, 2014.
36. Cristina Marcos, "GOP Chairman Bungles ObamaCare Math," *The Hill*, March 24, 2015.
37. Ezra Klein, "Health-Care Rashomon," *Washington Post*, January 19, 2011.
38. Dylan Scott, "Exclusive: Here's the House GOP's 'Incredibly Rigged' Obamacare Survey," Talking Points Memo, last modified May 1, 2014.
39. Kevin Drum, "For Republicans, Fear and Confusion Are All They Have Left," *Mother Jones*, last modified May 2, 2014.
40. Robert Pear, "Called by Republicans, Health Insurers Deliver Unexpected Testimony," *New York Times*, May 7, 2014.
41. Elise Viebeck, "GOP Struggles to Land Punches at Obamacare Insurance Hearing," *The Hill*, May 7, 2014.
42. Jonathan Topaz, "Carson Draws Raucous Applause," Politico, last modified January 24, 2015.
43. Donald J. Trump, interview by Dana Bash, CNN, July 29, 2015.
44. Donald J. Trump, interview by Scott Pelley, *60 Minutes*, CBS, September 27, 2015.
45. Robert Pear and Maggie Haberman, "Donald Trump's Health Care Ideas Bewilder Republican Experts," *New York Times*, April 8, 2016.

46. Craig Gilbert, "No Quick Obamacare Replacement, Ryan Says," *Milwaukee Journal Sentinel*, December 5, 2016.
47. Scott Pelley, "Speaker Ryan's 'Strange Bedfellows' Partnership with Trump," CBS, *60 Minutes*, December 4, 2016.
48. Jordan Weissmann, "I Can't Even with Paul Ryan and Health Care Anymore," Slate, last modified January 19, 2017.
49. Robert Costa and Amy Goldstein, "Trump Vows 'Insurance for Everybody' in Obamacare Replacement Plan," *Washington Post*, January 15, 2017.
50. "Donald Trump's News Conference: Full Transcript and Video," *New York Times*, last modified January 11, 2017.
51. David M. Drucker, "Republicans plot how to avoid their own midterm meltdown," *Washington Examiner*, January 5, 2017.
52. Maggie Haberman and Robert Pear, "Trump Tells Congress to Repeal and Replace Health Care Law 'Very Quickly,'" *New York Times*, January 10, 2017.
53. John Harwood, "Congressional Republicans Don't Expect Trump to Offer His Own Health or Tax Plans," CNBC online, last modified February 22, 2017.
54. Paul McLeod, "GOP Congressman Says a Drop in the Number of Insured People Could Be 'A Good Thing,'" BuzzFeed, last modified February 23, 2017.
55. White House, "Remarks by President Trump in Meeting with the National Governors Association," news release, February 27, 2017.
56. Avik Roy, "GOP's Obamacare Replacement Will Make Coverage Unaffordable for Millions—Otherwise, It's Great," *Forbes*, March 7, 2017.
57. Aviva Aron-Dine and Tara Straw, *House Tax Credits Would Make Health Insurance Far Less Affordable in High-Cost States* (Washington, DC: Center on Budget and Policy Priorities, March 16, 2017).
58. John Harwood, "The Republican Health-Care Bill Threatens Trump's Voters," CNBC online, last modified March 9, 2017.
59. Peter Suderman, "The GOP's Obamacare Repeal Bill Is Here. Is This Just Obamacare Lite?," *Reason*, last modified March 6, 2017.
60. Ezra Klein, "The GOP Health Bill Doesn't Know What Problem It's Trying to Solve," Vox, last modified March 6, 2017.
61. David Lazarus, "'Cheaper, better 'insurance for everybody'? Good luck with that," *Los Angeles Times*, January 20, 2017.
62. White House, "Press Briefing by Press Secretary Sean Spicer," news release, March 7, 2017.

63. Chris Murphy, United States Senator for Connecticut, "Senate Democrats: Passage of Affordable Care Act Was Open and Transparent, with Hundreds of Hours of Public Hearings and Dozens of Committee Meetings—After Passage of Law, Health Care Costs and Insurance Numbers Are Proof That ACA Works," press release, December 9, 2014.
64. Leonard Lance, "GOP Rep.: We Aren't Rushing Trumpcare," interview by Chris Hayes, *All In*, MSNBC, March 9, 2017.
65. Olga Khazan, "The Biggest Criticism of Paul Ryan's Health-Care Pitch," *The Atlantic*, last modified March 10, 2017.
66. *American Health Care Act* (Washington, DC: Congressional Budget Office, March 13, 2017).
67. Amanda Terkel, "Trump Worked a Few Weeks on Health Care. White House Claims He Gave It 'His All,'" HuffPost, last modified March 24, 2017.
68. Matt Shuham, "Trump on GOP Health Care Plan: 'Actually It's Very Simple,'" Talking Points Memo, last modified March 7, 2017.
69. Josh Dawsey, "Trump Lets His Aides Sweat the Details on Health Care," Politico, last modified March 20, 2017.
70. Tim Alberta, "Inside the GOP's Health Care Debacle," Politico, last modified March 24, 2017.
71. Manu Raju (@mkraju), "TRUMP warns House GOP of "political problems" if health care fails. "No details" on policy, laments Rep. Walter Jones, who is still a 'no,'" Twitter, March 21, 2017, 9:57 A.M.
72. Ryan Lizza (@RyanLizza), "This is a nice overview of where the healthcare debate stands from the perspective of the Freedom Caucus. (Note the comment about Trump.)," Twitter, March 22, 2017, 9:24 P.M.
73. Reuters, "Ryan Opposes Trump Working with Democrats on Healthcare," last modified March 29, 2017.
74. Noam N. Levey, "GOP Shuts Out Doctors, Experts, Democrats—Pretty Much Everybody—as They Work on Obamacare Repeal," Noam N. Levey, *Los Angeles Times*, April 27, 2017.
75. Billy House and Anna Edgerton, "GOP Tries Last-Minute Fix to Prevent Health Bill Collapse," Bloomberg, last modified May 2, 2017.
76. Alice Ollstein, "House GOPer: Move to Another State If You Have A Pre-Existing Condition," Talking Points Memo, last modified May 2, 2017.
77. Paul Kane, "Republicans Didn't Like Their Health-Care Bill but Voted for It Anyway," *Washington Post*, May 4, 2017.

78. Jerry Zremski, "Watch: Chris Collins Admits He Didn't Read Health Care Bill," *Buffalo News*, May 4, 2017.
79. Kane, "Republicans Didn't Like Their Health-Care Bill."
80. Esme Cribb, "GOP Rep. Who Voted for ACA Repeal Says It's 'Not the Bill We Promised' Voters," Talking Points Memo, last modified May 7, 2017.
81. Noam N. Levey and Lisa Mascaro, "Republican Secrecy Faces Mounting Criticism as GOP Senators Work Behind Closed Doors to Replace Obamacare," *Los Angeles Times*, June 16, 2017.
82. Jennifer Bendery, "GOP Senator Slams His Party's Process for Crafting Health Care Bill," HuffPost, last modified May 23, 2017.
83. Tara Golshan, Dylan Scott, and Jeff Stein, "We Asked 8 Senate Republicans to Explain What Their Health Bill Is Trying to Do," Vox, last modified June 16, 2017.
84. Alice Ollstein, "After CBO Report, Rand Paul Says He Will Oppose 'Terrible' O'care Repeal Bill," Talking Points Memo, last modified June 26, 2017.
85. Glenn Thrush and Jonathan Martin, "On Senate Health Bill, Trump Falters in the Closer's Role," *New York Times*, June 27, 2017.
86. Andrew Egger, "Trump Rallies Senate Support for Health Care," *Washington Examiner*, last modified June 26, 2017.
87. Robert Costa, Sean Sullivan, Juliet Eilperin, and Kelsey Snell, "How the Push for a Senate Health-Care Vote Fell Apart Amid GOP Tensions," *Washington Post*, June 28, 2017.
88. Jonathan Martin (@jmartNYT), "GOP senator calls just now, sez Trump consumed w RUSSIA Also: Trump must sell tax reform 'His vocabulary on healthcare was bout 10 words,'" Twitter, August 22, 2017, 6:20 P.M.
89. Zeke J. Miller, "Read Donald Trump's Interview with *Time* on Being President," *Time*, May 11, 2017.
90. "Scaramucci Talks Trump's Agenda, Relationship with the Press," *Fox News Sunday*, Fox News, last modified July 23, 2017.
91. NBC Politics (@NBCPolitics), "Senate Intel Committee Chairman Richard Burr on health care: 'I'll vote for anything,'" Twitter, July 24, 2017, 5:46 P.M.
92. Harry Stein (@HarrySteinDC). "GRAHAM: Skinny repeal is a 'fraud' and 'disaster.,'" Twitter, July 27, 2017, 5:20 P.M.
93. Ross Douthat (@DouthatNYT), "But to vote 'yes' under these circumstances - dead of night vote for a bill with no scrutiny that nobody wants - is just disgraceful.," Twitter, July 27, 2017, 10:05 P.M.
94. Senator Chris Murphy (@ChrisMurphyCT), "Seriously, this is weapons

grade bonkers. 3 Senators just announced they will vote for repeal only if assured it will never become law.," Twitter, July 27, 2017, 5:45 P.M.

95. Brett LoGiurato, "Trump fumes as McConnell says it's 'time to move on' after healthcare collapse," Business Insider, July 28, 2017.
96. Caitlin Owens and Sam Baker, "Repeal First, Ask Questions Later," Axios, last modified September 20, 2017.
97. Jason Noble, "Grassley: Fulfilling Campaign Promise Just as Important as 'Substance' of Health Bill," *Des Moines (IA) Register*, September 20, 2017.
98. Caitlin Owens and Sam Baker, "Repeal First, Ask Questions Later," Axios, last modified September 20, 2017.
99. Philip Klein, "Republicans Will Be Haunted by Their Desperate Defense of Obamacare's Pre-Existing Condition Ban," *Washington Examiner*, last modified October 4, 2018.
100. David Morgan, "Republicans Prepare for 'Obamacare' Showdown, with Eye to 2014 Elections," Reuters, last modified July 25, 2013.
101. Elise Viebeck, "GOP to Constituents: Questions on ObamaCare? Call Obama," *The Hill*, June 15, 2013.
102. Niels Lesniewski, "McConnell, Cornyn Urge NFL, Sports Leagues Not to Promote Obamacare," *Roll Call*, last modified June 28, 2013.
103. Todd S. Purdum, "The Obamacare Sabotage Campaign," Politico, last modified November 1, 2013.
104. Amy Goldstein and Juliet Eilperin, "HealthCare.gov: How Political Fear Was Pitted Against Technical Needs," *Washington Post*, November 2, 2013.
105. Haeyoun Park and Margot Sanger-Katz, "We're Tracking the Ways Trump Is Scaling Back Obamacare. Here Are 14," *New York Times*, July 11, 2018.
106. Sam Baker, "The Trump Effect on ACA Premiums," Axios, last modified May 24, 2018.
107. Michelle Ye Hee Lee and Glenn Kessler, "President Trump's Biggest Obamacare Bloopers," *Washington Post*, March 22, 2017.
108. Steve Benen, "Trump's curious boast: he 'decided not to' kill the ACA," MSNBC online, last modified June 25, 2019.

CHAPTER 4: "Extending a Middle Finger to the World": Climate Change and Energy Policy

1. Daniel R. Coats, *Statement for the Record: Worldwide Threat Assessment of the US Intelligence Community* (Washington, DC: Office of the Director of National Intelligence, January 29, 2019).

2. Deborah Zabarenko, "Is U.S. near a tipping point on global warming?" Reuters, February 19, 2007.
3. Coral Davenport and Eric Lipton, "How G.O.P. Leaders Came to View Climate Change as Fake Science," *New York Times*, June 3, 2017.
4. Kate Sheppard, "McCain Now Says Climate Science Might Be 'Flawed,'" *Mother Jones*, last modified October 12, 2010.
5. Eric Holthaus, "Senate Votes 98-1 That Climate Change Is Real but Splits on That Pesky Cause," Slate, January 21, 2015.
6. Kevin Drum, "Climate Change Goes Back to Square Zero," *Mother Jones*, last modified January 27, 2012.
7. David Weigel, "Yes, Mike Huckabee Backed Cap and Trade in 2007," Slate, last modified December 15, 2010.
8. Darren Samuelsohn, "Pawlenty Can't Outrun Climate Past," Politico, last modified March 22, 2011.
9. Brad Plumer, "That Time Marco Rubio Called for a 'Cap-and-Trade or Carbon Tax Program,'" Vox, last modified January 13, 2016.
10. Brad Johnson, "Supported by Tea Party Polluters, Upton Flips on Threat of Global Warming," ThinkProgress, last modified December 28, 2010.
11. Eric Kleefeld, "Pawlenty: My Past Support for Cap-and-Trade 'Was Stupid,'" ThinkProgress, last modified March 29, 2011.
12. Kate Sheppard, "House Republicans Bring Strange Theories and Wacky Witnesses to Climate Hearings," Grist, last modified April 20, 2009.
13. Ursula Goodenough, "Is the Good Book a Good Guide in the Climate Change Debate?," National Public Radio, last modified November 18, 2010.
14. Satyam Khanna, "Rep. Barton: Climate Change Is 'Natural,' Humans Should Just 'Get Shade,'" ThinkProgress, last modified March 26, 2009.
15. Lee Fang, "Rep. Akin Argues Against Curbing Emissions: I Don't Want to Stop the Seasons from Changing," ThinkProgress, last modified June 3, 2009.
16. Lee Fang, "Member of Congressional Science Committee: Global Warming a 'Fraud' to 'Create Global Government,'" *Nation*, last modified August 10, 2013.
17. Kate Sheppard, "Which Dems Are Backing Murkowski's Attack on Carbon Regs?," *Mother Jones*, last modified June 9, 2010.
18. Matt Corley, "Boehner: It's 'Almost Comical' to Say Carbon Dioxide and Climate Change Are Dangerous Since Cows Fart a Lot," ThinkProgress, last modified April 20, 2009.

19. Brian Beutler, "MIT Scientist: Republicans Misusing My Climate Change Paper," Talking Points Memo, last modified April 2, 2009.
20. Ezra Klein, "How Much Will Cap-and-Trade Cost You?," *Washington Post*, June 22, 2009.
21. Ben Geman, "GOP Ex-Lawmaker: Facts Will 'Overwhelm' GOP Opposition to Climate Change," *The Hill*, July 23, 2012.
22. Ben Geman, "Republicans Vote to Thwart Imaginary Drilling Restriction," *National Journal*, July 11, 2014.
23. Pete Kasperowicz, "GOP: Predict Storms, Not Climate Change," *The Hill*, March 28, 2014.
24. Ali Watkins, "Republicans Hit CIA on . . . Climate Change?," HuffPost, last modified March 17, 2015.
25. Rebecca Leber, "House Republicans Use "Energy Independence" as an Excuse to Slash Climate Funds," *New Republic* online, last modified March 17, 2015.
26. Marina Koren, "Ted Cruz Tells NASA to Stop Worrying About Climate Change and Focus on Space," *National Journal*, March 12, 2015.
27. David Roberts, "This Congressman Doesn't Want a Federal Science Board to Be Allowed to Consider Science," Grist, last modified March 4, 2015.
28. Roberts, "This Congressman."
29. "The Senate, a Snowball, and a Climate Change Skeptic," NBC News online, last modified February 26, 2015.
30. "Video: One-on-One Interview with Ted Cruz," Jay Root and Todd Wiseman, Texas Tribune, last modified March 24, 2015.
31. Philip Bump, "Ted Cruz Compares Climate Change Activists to 'Flat-Earthers.' Where to Begin?," *Washington Post*, March 25, 2015.
32. Aliyah Frumin, "John Boehner: I'm 'Not Qualified' to Debate Climate Change," MSNBC online, last modified May 29, 2014.
33. Marc Caputo, "Rick Scott Won't Say If He Thinks Man-Made Climate-Change Is Real, Significant," *Miami Herald*, May. 27, 2014.
34. Joseph Gerth, "McConnell Won't Tackle Climate Change," *Louisville (K.Y.) Courier Journal*, October 2, 2014.
35. Rebecca Leber, "Someone Tell Bobby Jindal You Don't Need to Be a Scientist to Understand Science," *New Republic* online, last modified September 16, 2014.
36. Dylan Matthews, "Donald Trump has tweeted climate change skepticism 115 times. Here's all of it," Vox, June 1, 2017.

37. Ben Schreckinger, "Trump Acknowledges Climate Change—at His Golf Course," Politico, last modified May 23, 2016.
38. Aaron Blake, "The First Trump-Clinton Presidential Debate Transcript, Annotated," *Washington Post*, September 26, 2016.
39. Sam Stein, "There Is No Good Way to Explain Donald Trump's Climate Change Tweet," HuffPost, last modified September 27, 2016.
40. Isaac Arnsdorf, Josh Dawsey, and Seung Min Kim, "Trump's Flashy Executive Actions Could Run Aground," Politico, last modified January 25, 2017.
41. Joshua Green, "Why the Fight over the Keystone Pipeline Is Completely Divorced from Reality," Bloomberg, last modified January 7, 2015.
42. Mike Allen, "Scott Pruitt's Laundry List of Scandals While with the EPA," Axios, last modified June 15, 2018.
43. Coral Davenport and Lisa Friedman, "In His Haste to Roll Back Rules, Scott Pruitt, E.P.A. Chief, Risks His Agenda," *New York Times*, April 7, 2018.
44. Michael Grunwald, "The Myth of Scott Pruitt's EPA Rollback," Politico, last modified April 7, 2018.
45. Ashley Parker, Philip Rucker, and Michael Birnbaum, "Inside Trump's Climate Decision: After Fiery Debate, He 'Stayed Where He's Always Been,'" *Washington Post*, June 1, 2017.
46. Alex Johnson, Erik Ortiz, and Becky Bratu, "Casino Robbery Ends with Dozens Dead at Resort in Philippines," NBC News online, last modified June 1, 2017.
47. NBC Nightly News with Lester Holt (@NBCNightlyNews), "US intel official: Pres. Trump 'was freelancing' with the terrorism declaration and 'a laugh went up in the Situation Room' when he made it.," Twitter, June 1, 2017, 1:43 P.M.
48. David Roberts, "The 5 Biggest Deceptions in Trump's Paris Climate Speech," Vox, last modified June 2, 2017.
49. Michael Grunwald, "Why Trump Actually Pulled Out of Paris," Politico, last modified June 1, 2017.
50. Matt Shuham, "Pence: 'For Some Reason or Another,' the Left Cares About Climate Change," Talking Points Memo, last modified June 2, 2017.
51. Linda Qiu and John Schwartz, "Trump's False Claims About Coal, the Environment, and West Virginia," *New York Times*, August 21, 2018.
52. Associated Press, "Read the Transcript of AP's Interview with President Trump," last modified October 17, 2018.

53. Joel Shannon, "'I Know a Lot About Wind,' Trump Says. A Government FAQ Proves He Doesn't," *USA Today*, March 29, 2019.
54. Ledyard King, "Do Wind Farms Cause Cancer? Some Claims Trump Made About the Industry Are Just Hot Air," *USA Today*, April 3, 2019.
55. Brad Plumer, "We Fact-Checked President Trump's Dubious Claims on the Perils of Wind Power," *New York Times*, April 3, 2019.
56. David Knowles, "Trump Just a Blowhard on Windmills, Lawmakers Say of 'Idiotic' Comments," Yahoo! News, last modified April 4, 2019.
57. "Scottish Government Wins Donald Trump Wind Power Legal Costs," BBC online, last modified February 28, 2019.
58. *Report on Effects of a Changing Climate to the Department of Defense* (Washington, DC: Office of the Under Secretary of Defense for Acquisition and Sustainment, Department of Defense, January 2019).
59. Donald J. Trump (@RealDonaldTrump), "In the beautiful Midwest, windchill temperatures are reaching minus 60 degrees, the coldest ever recorded. In coming days, expected to get even colder. People can't last outside even for minutes. What the hell is going on with Global Waming? Please come back fast, we need you!," Twitter, January 28, 2019, 9:28 P.M.
60. Colby Itkowitz, "'Are You Serious?': John Kerry's Climate Change Credentials Challenged by GOP Lawmaker," *Washington Post*, April 10, 2019.
61. Tim Dickinson, "Is This the Dumbest Moment in Congressional History?," *Rolling Stone online*, last modified April 10, 2019.
62. John Kerry (@JohnKerry), "It's almost as if someone said 'Congress has hit rock bottom' and Massie replies with 'hold my beer.'" Twitter, April 10, 2019, 5:02 P.M.
63. Amy Harder, "GOP Lawmaker: Green New Deal Is Like Genocide," Axios, last modified March 14, 2019.
64. Shawn Langlois, "Ocasio-Cortez, Like Stalin, Is Coming for Your Burgers, Former Trump Aide Warns," MarketWatch, last modified March 1, 2019.
65. David Remnick, "Alexandria Ocasio-Cortez Is Coming for Your Hamburgers!," *New Yorker*, March 4, 2019.
66. Colby Itkowitz, "Sen. Mike Lee says we can solve climate change with more babies. Science says otherwise," *Washington Post*, March 26, 2019.
67. Jonathan Chait, "Republican Senator Mike Lee: Having Babies the Only Solution to Climate Change," *New York*, last modified March 26, 2019.
68. Roll Call Votes 116th Congress—1st Session, "On the Cloture Motion

(Motion to Invoke Cloture on the Motion to Proceed to H.R. 268)," U.S. Senate online, March 26, 2019.

CHAPTER 5: "A Series of Hasty Unplanned, Unexamined Decisions": Foreign Policy

1. Weekly Republican Radio Address, Senator Kit Bond (R-MO), May 8, 2009.
2. Randy Krehbiel, "Inhofe Blasts Obama at Grove Town Hall," *Tulsa World*, September 2, 2009.
3. Maj. Gen. Michael Lehnert (Retired), "I Helped Create Gitmo. Now I Want It Shut Down," Politico, last modified January 11, 2015.
4. Carol Rosenberg, "The Cost of Running Guantánamo Bay: $13 Million Per Prisoner," *New York Times*, September 16, 2019.
5. Christina Bellantoni, "Peter King Says Obama Could Make Country Safer by Using the Word Terrorism," Talking Points Memo, last modified January 6, 2010.
6. Lydia Wheeler, "Top GOP Senator: Call It a War on Terror," *The Hill*, January 11, 2015.
7. Lauren Fox, "GOP Senators: Paris Shooting Justifies NSA Powers," *National Journal*, January 7, 2015.
8. Donald J. Trump (@RealDonaldTrump), "Is President Obama going to finally mention the words radical Islamic terrorism? If he doesn't he should immediately resign in disgrace!," Twitter, June 12, 2016, 1:58 P.M.
9. Jonathan Weisman, Mark Landler, and Jeremy W. Peters, "As Obama Makes Case, Congress Is Divided on Campaign Against Militants," *New York Times*, September 8, 2014.
10. Mary Beth Sheridan, "START Expiration Ends U.S. Inspection of Russian Nuclear Bases," *Washington Post*, August 17, 2010.
11. David S. Broder, "Wanted: A Few Stand-up Candidates," *Washington Post*, August 19, 2010.
12. Kathy Lally, "Russia Wonders Why U.S. Would Turn Away from Treaty," *Washington Post*, November 19, 2010.
13. Mary Beth Sheridan and Walter Pincus, "Vote on New START Nuclear Arms Treaty Delayed in Senate," *Washington Post*, August 4, 2010.
14. Karen DeYoung, "Ex-Eastern Bloc Nations Tell U.S. Senate to Ratify New START," *Washington Post*, November 20, 2010.

15. Jonathan Chait, "Conservatives Hate the Iran Deal Because They Hate All Deals," *New York*, last modified March 29, 2015.
16. Associated Press, "Obama Offers GOP a Deal on Arms Treaty," last modified November 13, 2010.
17. Peter Baker, "G.O.P. Opposition Dims Hope for Arms Treaty with Russia," *New York Times*, November 16, 2010.
18. Roll Call Vote, 111th Congress, 2nd Session, U.S. Senate, December 22, 2010.
19. Michael Crowley, "Romney v. Obama Cont'd. Can the Center Hold?," *Time*, last modified July 8, 2010.
20. Julian Pecquet, "Sen. DeMint Taps Brakes on UN Treaty as Home-School Opposition Grows," *The Hill*, July 19, 2012.
21. Gail Collins, "Santorum Strikes Again," *New York Times*, December 5, 2012.
22. Michael Kranish, "A Lesson for Bob Dole: Old Rules No Longer Apply," *Boston Globe*, March 24, 2013.
23. Daniel W. Drezner, "Praised Be the Glorious Sovereigntists Who Protect the U.S.A. from . . . from . . . Wait, What?," *Foreign Policy*, last modified December 5, 2012.
24. Josh Rogin, "New Push Begins for Law of the Sea Treaty," *Foreign Policy*, last modified May 10, 2012.
25. Gavin Aronsen, "UN Adopts Historic Arms Trade Treaty Despite NRA Opposition," *Mother Jones*, last modified April 2, 2013.
26. "Kerry to Sign UN Arms Treaty, Despite Senators' Opposition," Fox News online, last modified September 24, 2013.
27. Jonathan Chait, "Dick Cheney's Ongoing Descent into Insanity Accidentally Clarifies Iran Debate," *New York*, last modified April 8, 2015.
28. Peter Sullivan, "Graham: Senate Will Vote on Iran Sanctions Legislation in January," *The Hill*, December 27, 2014.
29. S. V. Dáte, "47 GOP Senators Tell Iran They May Not Honor a Nuclear Deal," National Public Radio, last modified March 9, 2015.
30. Mary Ann Akers, "Democrats to Rep. Mark: Now Who's Un-American?," *Washington Post*, June 11, 2009.
31. David Weidner, interview by Rachel Maddow, *The Rachel Maddow Show*, MSNBC, June 12, 2009.
32. "Editorial: GOP Letter to Iran Disgraces America," *Detroit Free Press*, March 9, 2015.

33. David Goldstein, "Precedent for GOP's Iran Letter Hard to Find, Historians Say," McClatchy, last modified March 10, 2015.
34. Burgess Everett, "Cotton Storms the Senate," Politico, last modified March 10, 2015.
35. American Bridge (@merican_Bridge), ".@SenRandPaul got on TV this morning and made exactly no sense about his consigning of the GOP letter to Iran: https://www.youtube.com/watch?v=yAVImGekhbw," Twitter, March 11, 2105, 10:23 A.M.
36. Alex Leary (@learyreports), ".@marcorubio PAC looks to raise money off Iran letter. 'Marco was proud to be one of the first senators to sign.,'" Twitter, March 11, 2015, 11:46 A.M.
37. "Rubio: Would 'Absolutely' Send Another Letter to Iran," NBC News online, last modified March 10, 2015.
38. Max Fisher, "This Is an Astoundingly Good Iran Deal," Vox, last modified July 16, 2015.
39. Niels Lesniewski, "GOP Hawks Quick to Blast Iran Framework," *Roll Call*, last modified April 2, 2015.
40. Dana Milbank, "Republicans' Knee-Jerk Hatred of the Iran Deal," *Washington Post*, July 14, 2015.
41. Jacob Heilbrunn, "Twilight of the Wise Man," *Foreign Policy* online, last modified October 12, 2011.
42. Michael Crowley, "Exclusive: Armitage to Back Clinton over Trump," Politico, last modified June 16, 2016.
43. Cooper Allen, "Brent Scowcroft Endorses Clinton over Trump," *USA Today*, June 22, 2016.
44. Michael Finnegan and Seema Mehta, "Trump Says He Opposed Iraq War from the Start. He Did Not," *Los Angeles Times*, October 9, 2016.
45. Philip Bump, "Donald Trump's Falsehood-Laden Press Conference, Annotated," *Washington Post*, July 27, 2016.
46. Jenna Johnson, "Donald Trump on Waterboarding: 'If It Doesn't Work, They Deserve It Anyway,'" *Washington Post*, November 23, 2015.
47. Eliza Collins, "Trump: I Consult Myself on Foreign Policy," Politico, last modified March 16, 2016.
48. "A Transcript of Donald Trump's Meeting with the *Washington Post* Editorial Board," *Washington Post*, March 21, 2016.
49. White House Office of the Press Secretary, "Press Conference by the President," Barack Obama, December 16, 2016.

50. Kristen Welker, Ken Dilanian, and Alexandra Jaffe, "Donald Trump Has Attended Only Two Intelligence Briefings," NBC News online, last modified November 24, 2016.
51. Louis Nelson, "Trump: I Don't Need Daily Briefings," Politico, last modified December 11, 2016.
52. Jim VandeHei and Mike Allen, "Reality Bites: Trump's Wake-up Call," Axios, last modified January 18, 2017.
53. David E. Sanger, Eric Schmitt, and Peter Baker, "Turmoil at the National Security Council, from the Top Down," *New York Times*, February 12, 2017.
54. Steve Holland and Jeff Mason, "Embroiled in Controversies, Trump Seeks Boost on Foreign Trip," Reuters, last modified May 17, 2017.
55. Robbie Gramer, "NATO Frantically Tries to Trump-Proof President's First Visit," *Foreign Policy*, last modified May 15, 2017.
56. Greg Jaffe and Philip Rucker, "National Security Adviser Attempts to Reconcile Trump's Competing Impulses on Afghanistan," *Washington Post*, August 4, 2017.
57. Matt Spetalnick, David Brunnstrom, and John Walcott, "Understanding Kim: Inside the U.S. Effort to Profile the Secretive North Korean Leader," Reuters, last modified April 26, 2018.
58. Patrick Radden Keefe, "McMaster and Commander," *New Yorker*, last modified April 23, 2018.
59. Greg Miller, "Gap Continues to Widen Between Trump and Intelligence Community on Key Issues," *Washington Post*, December 11, 2018.
60. John Walcott, "'Willful Ignorance.' Inside President Trump's Troubled Intelligence Briefings," *Time*, last modified February 2, 2019.
61. Daniel Lippman, "Trump's Diplomatic Learning Curve: Time Zones, 'Nambia' and 'Nipple,'" Politico, last modified August 13, 2018.
62. Jonathan Chait, "Trump Foreign Policy Held Back by Struggle to Grasp Time Zones, Maps," *New York*, last modified August 13, 2018.
63. Ayman Mohyeldin (@AymanM), "DHS official tells @NBCNews professional public servants at State/DHS had no input in drafting EO now scrambling to try and interpret them," Twitter, January 28, 2017, 3:20 P.M.
64. @AymanM, Twitter, January 28, 2017, 12:20 P.M.
65. John J. Harwood (@JohnJHarwood), "senior Justice official tells @NBCNews that Dept had no input. not sure who in WH is writing/reviewing. standard NSC process not functioning," Twitter, January 28, 2017, 12:45 P.M.

66. Eoghan Macguire, Ali Gostanian, and Erik Ortiz, "Trump Travel Restrictions Leave Refugees Stranded: Reports," NBC News online, last modified January 28, 2017.
67. Michael D. Shear, Nicholas Kulish, and Alan Feuer, "Judge Blocks Trump Order on Refugees Amid Chaos and Outcry Worldwide," *New York Times*, January 28, 2017.
68. Andrew Prokop, "In a Reversal, the Trump Administration Now Says Green Card Holders Can Enter the US," Vox, last modified January 29, 2017.
69. Rebecca Tan, "A Timeline of Trump's Clearly Made-up 'Secret Plan' to Fight ISIS," Vox, last modified July 3, 2017.
70. Ben Schreckinger, "Trump Would Turn to Generals for Islamic State Plan," Politico, last modified September 6, 2016.
71. Cynthia McFadden and William M. Arkin, "Trump's Pentagon Presents ISIS Plan That Looks Much Like Obama's," NBC News online, last modified March 17, 2017.
72. Kimberly Dozier, "U.S. Commandos Running Out of ISIS Targets," Daily Beast, last modified July 5, 2017.
73. Gerard Baker, Carol E. Lee, and Michael C. Bender, "Trump Says He Offered China Better Trade Terms in Exchange for Help on North Korea," *Wall Street Journal*, April 12, 2017.
74. Aaron Blake, "President Trump's Thoroughly Confusing Fox Business Interview, Annotated," *Washington Post*, April 12, 2017.
75. Anna Fifield, "North Korea Taps GOP Analysts to Better Understand Trump and His Messages," *Washington Post*, September 26, 2017.
76. Mark Landler, "North Korea Asks for Direct Nuclear Talks, and Trump Agrees," *New York Times*, March 8, 2018.
77. Kevin Liptak and Jeremy Diamond, "From Grin to Grim: Inside the Trump-Kim Summit Collapse," CNN online, last modified May 24, 2018.
78. Brian Bennett and Tessa Berenson, "President Trump 'Doesn't Think He Needs' to Prepare Much for His Meeting with North Korea's Kim Jong Un," *Time*, last modified May 16, 2018.
79. Nicholas Kristof, "Trump Was Outfoxed in Singapore," *New York Times*, June 12, 2018.
80. Salvador Rizzo, "Five Whoppers from President Trump's Impromptu News Conference," *Washington Post*, June 15, 2018.
81. Donald J. Trump (@RealDonaldTrump), "Before taking office people were assuming that we were going to War with North Korea. President Obama

said that North Korea was our biggest and most dangerous problem. No longer—sleep well tonight!," Twitter, June 13, 2018, 6:01 A.M.

82. Sharon Shi and Clement Burge, "While Trump and Kim Talk, North Korea Appears to Expand Its Nuclear Arsenal, *Wall Street Journal*, July 27, 2019.
83. Daniel Dale, "Trump Makes 13 False Claims in Cabinet Meeting," CNN online, last modified July 16, 2019.
84. John Hudson and Josh Dawsey, "Trump Botches North Korea Sanctions Announcement, Sparking Widespread Confusion," *Washington Post*, March 22, 2019.
85. Andrew Restuccia and Caitlin Oprysko, "Trump Surprises His Own Aides by Reversing North Korea Sanctions," Politico, last modified March 22, 2019.
86. Zachary Cohen (@ZcohenCNN), "More than two hours later, me & @JDiamond1 are told that administration officials are still waiting for guidance from the White House on the meaning of Trump's tweet and how to proceed. For now, they are simply in a holding pattern.," Twitter, March 22, 2019, 4:20 P.M.
87. Saleha Mohsin, Jennifer Jacobs, and Nick Wadhams, "Trump Tried to Undo North Korea Penalty, Contrary to U.S. Account," Bloomberg, last modified March 26, 2019.
88. Richard N. Haass (@RichardHaass), "in just 24 hours the secstate was surprised by Golan Heights tweet & now @realDonaldTrump has lifted new Treasury sanctions vs NK b/c he likes Chairman Kim. There is not even the pretense of a national security process. Hard to imagine what would occur if there were a real crisis," Twitter, March 22, 2019, 2:30 P.M.
89. Peter Baker, "Trump in Asia: 'We're the Hottest Show in Town,'" *New York Times*, July 1, 2019.
90. Anita Kumar, "Trump's Made-for-TV Moment in North Korea," Politico, last modified June 30, 2019.
91. Senator Brian Schatz (@brianschatz), "He's just not good at negotiating, actually. He just gave away something important and he either doesn't know or doesn't care.," Twitter, June 30, 2019, 1:58 P.M.
92. Andrea Mitchell (@mitchellreports), "Right now @MSNBC live NBC Korea expert Victor Cha says meeting Kim at DMZ is like having the Super Bowl when you haven't played a single game. This meeting should be at the end of successful talks not when there's been no progress," Twitter, June 30, 2019, 1:11 A.M.

93. White House, "Remarks by President Trump and Crown Prince Salman of Bahrain Before Bilateral Meeting," news release, September 16, 2019.
94. Jeremy Diamond, "Trump Suggests U.S. 'Dumb Son of a Bitch' on Iran Deal," CNN online, last modified December 17, 2015.
95. Carol Morello, "Iran Nuclear Deal Could Collapse Under Trump," *Washington Post*, November 9, 2016.
96. Alexandra Jaffe, "Donald Trump: 'Very Hard to Say, "We're Ripping Up"' Iran Deal," NBC News online, last modified August 16, 2015.
97. Matthew Lee, "Trump Administration: Iran Complying with Nuclear Deal," *USA Today*, April 19, 2017.
98. Peter Baker, "Trump Recertifies Iran Nuclear Deal, but Only Reluctantly," *New York Times*, July 17, 2017.
99. David E. Sanger, "Trump Seeks Way to Declare Iran in Violation of Nuclear Deal," *New York Times*, July 27, 2017.
100. Karen DeYoung, "Allies Fume over Trump's Withdrawal from Iran Deal but Have Few Options to Respond," *Washington Post*, May 14, 2018.
101. Ronen Bergman and Mark Mazzetti, "The Secret History of the Push to Strike Iran," *New York Times Magazine*, September 4, 2019.
102. Jeva Lange, "Trump Announces That the U.S. Will Withdraw from the Iran Deal," *The Week* online, last modified May 18, 2018.
103. Laura Silver, "Trump Has Told Iran Not To Threaten The U.S. And, Oh Man, He's Got The Caps Lock On," BuzzFeed, July 23, 2018.
104. Julian Borger and Bethan McKernan, "Trump: I'll Be Iran's 'Best Friend' If It Acquires No Nuclear Weapons," *Guardian* (US edition), last modified June 22, 2019.
105. Joshua Keating, "Jared's Not So Sure He Can Bring Peace to the Middle East. Thank God," Slate, August 1, 2017.
106. Reuters, "Highlights of Reuters Interview with Trump," last modified April 27, 2017.
107. David A. Graham, "Trump: Middle East Peace Is 'Not as Difficult as People Have Thought,'" *The Atlantic*, May 3, 2017.
108. Ashley Feinberg, "Kushner on Middle East Peace: 'What Do We Offer That's Unique? I Don't Know,'" *Wired*, last modified August 1, 2017.
109. Josh Dawsey, Missy Ryan, and Karen DeYoung, "Trump Had for Months Been Determined to Move U.S. Embassy to Jerusalem," *Washington Post*, December 6, 2017.
110. Dawsey, Ryan, and DeYoung, "Trump Had for Months."

111. Mark Landler, "For Trump, an Embassy in Jerusalem Is a Political Decision, Not a Diplomatic One," *New York Times*, December 6, 2017.
112. Joshua Keating, "Trump's Bizarre Belief That Taking Jerusalem 'Off the Table' Makes Peace More Likely," Slate, last modified January 25, 2018.
113. Dominique Mosbergen, "Trump Says He Made Major Israel Decision After Quick 'Little History' Lesson," HuffPost, last modified April 7, 2019.
114. Anne Applebaum, "Trump Has the Attention Span of a Gnat. It's Destroying Our Foreign Policy," *Washington Post*, May 10, 2019.
115. Peter Baker, Mujib Mashal, and Michael Crowley, "How Trump's Plan to Secretly Meet with the Taliban Came Together, and Fell Apart," *New York Times*, September 8, 2019.
116. Justin Wise, "Dem Lawmaker: Who the 'F–' Thought It Was a Good Idea to Invite Taliban to Camp David?," *The Hill*, September 9, 2019.
117. White House, "Remarks by President Trump Before Marine One Departure," news release, September 9, 2019.
118. Susan B. Glasser, "Trump Is Finally Rid of John Bolton, but Does It Really Matter?," *New Yorker*, September 11, 2019.

CHAPTER 6: "The Cruelty Is the Point": The Collapse of Immigration Policy

1. Ron Brownstein, "The Immigration Impasse," *National Journal*, May 1, 2010.
2. "Obama's Remarks on Immigration," *New York Times*, May 10, 2011.
3. "Election 2012: President Exit Polls," *New York Times*, accessed November 9, 2012.
4. John Parkinson, "Boehner: Raising Tax Rates 'Unacceptable,'" ABC News online, last modified November 8, 2012.
5. Michael O'Brien, "GOP Resistance to Immigration Reform Could Be Casualty of 2012 Election," NBC News online, last modified November 9, 2012.
6. Matthew Kaminski, "Marco Rubio: Riding to the Immigration Rescue," *Wall Street Journal*, January 14, 2013.
7. Adam Serwer, "The Rubio Immigration Plan Conservatives Love Looks a Lot Like Obama's," *Mother Jones*, last modified January 15, 2013.
8. Sahil Kapur, "McCain: Improved Border Situation Has Made Immigration Reform Possible," Talking Points Memo, last modified January 29, 2013.
9. Ryan Lizza, "Getting to Maybe," *New Yorker*, June 24, 2013.

10. *The Economic Impact of S. 744, the Border Security, Economic Opportunity, and Immigration Modernization Act* (Washington, DC: Congressional Budget Office, June 2013).
11. "Transcript: Exclusive Interview with House Speaker John Boehner on NSA Leak, Immigration Reform, and More," ABC News online, last modified June 11, 2013.
12. Ginger Gibson, "Boehner: No Vote on Senate Immigration Bill," Politico, modified July 8, 2013.
13. Greg Sargent, "Syria Won't Make GOP's Immigration Problem Go 'Poof" and Disappear," *Washington Post*, September 12, 2013.
14. "Hey Mr. Speaker, Find Time to Read the Bill!," MSNBC online, last modified November 14, 2013.
15. Kristina Peterson and John D. McKinnon, "Health-Law Fracas Leaves Congress in Limbo," *Wall Street Journal*, November 17, 2013.
16. "Boehner: No Immigration Reform Until Obama Regains Our Trust," NBC News online, last modified February 6, 2014.
17. Jennifer Epstein, "Cantor: Obama Must Build on Immigration," Politico, last modified February 2, 2014.
18. Greg Sargent, "Morning Plum: GOP May Kill Immigration Reform Because #OBUMMER," *Washington Post*, February 3, 2014.
19. Suzy Khimm, "Schumer: Start Immigration Reform in 2017," MSNBC online, last modified February 9, 2014.
20. Jonathan Cohn, "Calling the Republican Bluff on Immigration," *New Republic*, last modified February 9, 2014.
21. Tim Devaney, "Rubio: We'll 'Never Have the Votes' for Immigration Reform Until Border Secured," *The Hill*, August 3, 2014.
22. Sahil Kapur, "John McCain Throws In the Towel on Immigration Reform Bill," Talking Points Memo, last modified August 25, 2014.
23. Aaron Blake, "The National GOP Just Labeled Marco Rubio's Immigration Bill 'Amnesty,'" *Washington Post*, September 9, 2014.
24. Ezra Klein, "Obama's Advantage Is That He Has an Immigration Policy. Republicans Don't," Vox, last modified November 20, 2014.
25. Jerry Markon, "Fewer Immigrants Are Entering the U.S. Illegally, and That's Changed the Border Security Debate," *Washington Post*, May 27, 2015.
26. Joshua Green, *Devil's Bargain: Steve Bannon, Donald Trump, and the Storming of the Presidency* (New York: Penguin Press, 2017).

27. "Editorial: A Chance to Reset the Republican Race," *New York Times*, January 30, 2016.
28. Jenna Johnson, "'Build That Wall' Has Taken On a Life of Its Own at Donald Trump's Rallies—but He's Still Serious," *Washington Post*, February 12, 2016.
29. Bob Woodward and Robert Costa, "Trump Reveals How He Would Force Mexico to Pay for Border Wall," *Washington Post*, April 5, 2016.
30. Joshua Green, "How to Get Trump Elected When He's Wrecking Everything You Built," Bloomberg, last modified May 26, 2016.
31. Mallory Shelbourne, "Trump Wants to Keep New Immigrants from Getting Welfare—Which Is Already Law," *The Hill*, June 21, 2017.
32. Michael D. Shear, "White House Makes Hard-Line Demands for Any 'Dreamers' Deal," *New York Times*, October 8, 2017.
33. White House, "Remarks by President Trump and NATO Secretary General Jens Stoltenberg Before Bilateral Meeting," news release, April 2, 2019.
34. White House, "Remarks by President Trump in Cabinet Meeting," news release, February 12, 2019.
35. "2018 National Drug Threat Assessment," U.S. Department of Justice Drug Enforcement Administration, October 2018.
36. "President Trump Went to a Border Town to Prove They Need a Wall. Residents Say Otherwise," Gina Martinez, *Time*, January 10, 2019.
37. Dara Lind, "Trump Claimed Women Were Gagged with Tape. Then Border Patrol Tried to Find Some Evidence," Vox, last modified January 27, 2019.
38. Jill Castellano and Rafael Carranza, "Behind Efforts to Build Prototypes of Trump's Border Wall, Emails Show a Confusing and Haphazard Process," *USA Today*, October 22, 2017.
39. Donald J. Trump (@RealDonald Trump), "The Wall is the Wall, it has never changed or evolved from the first day I conceived of it. Parts will be, of necessity, see through and it was never intended to be built in areas where there is natural protection such as mountains, wastelands or tough rivers or water.....," Twitter, January 18, 2018, 6:15 A.M.
40. Philip Bump, "How Trump's Wall Has Evolved (Despite His Denials That It Has)," *Washington Post*, January 18, 2018.
41. Associated Press, "Trump's Wall Prototypes to Come Down Along U.S.-Mexico Border," NBC News online, last modified February 22, 2019.
42. Donald J. Trump (@RealDonald Trump), "A design of our Steel Slat Barrier

which is totally effective while at the same time beautiful!,"Twitter, December 21, 2018, 5:14 P.M.

43. Rebecca Shabad, Garrett Haake, Leigh Ann Caldwell, and Hallie Jackson, "Partial Government Shutdown to Continue Through Next Week," NBC News online, last modified December 22, 2018.
44. "Gov. John Kasich 'Seriously Look at' Running for President in 2020, Says Dysfunction in Washington Is 'Very Disturbing,'" Fox News Sunday, Fox News online, last modified December 23, 2018.
45. Nick Miroff and Josh Dawsey, "Trump Wants His Border Barrier to Be Painted Black with Spikes. He Has Other Ideas, Too," *Washington Post*, May 16, 2019.
46. Anna Giaritelli, "Trump Has Not Built a Single Mile of New Border Fence After 30 Months in Office," *Washington Examiner*, last modified July 20, 2019.
47. Giaritelli, "Trump Has Not Built."
48. Denise Lu, "The Border Wall: What Has Trump Built So Far?," *New York Times*, February 12, 2019.
49. Steve Benen, "The Unsettling List of Trump's Made-up Conversations Keeps Growing," MSNBC online, last modified January 4, 2019.
50. D'Angelo Gore and Eugene Kiely, "Fact Check: Trump's Border Boast," *USA Today*, July 31, 2017.
51. Donald J. Trump (@RealDonaldTrump), "....however, for strictly political reasons and because they have been pulled so far left, do NOT want Border Security. They want Open Borders for anyone to come in. This brings large scale crime and disease. Our Southern Border is now Secure and will remain that way.......," Twitter, December 11, 2018, 7:12 A.M.
52. Philip Bump, "Trump's Arguments for Necessity of Border Wall Have Already Been Broadly Debunked," *Washington Post*, December 11, 2018.
53. "'Hardball with Chris Matthews' for Tuesday, August 18, 2015," *Hardball with Chris Matthews*, MSNBC online, last modified August 18, 2015.
54. Michael Scherer, "2016 Person of the Year Donald Trump," *Time*, December 7, 2016.
55. Associated Press, "'Dreamers' Should 'Rest Easy,' Trump says," last modified April 21, 2017.
56. Michael D. Shear and Julie Hirschfeld Davis, "Trump Moves to End DACA and Calls on Congress to Act," *New York Times*, September 5, 2017.

57. Caitlin Mac Neal, "Trump on DACA Deal with Congress: 'The Wall Will Come Later,'" Talking Points Memo, last modified September 14, 2017.
58. NBC Politics (@NBCPolitics), "Watch President Trump and congressional leaders debate immigration policy," Twitter, January 9, 2018, 1:52 P.M.
59. Leigh Ann Caldwell, "Bipartisan Senate Group Finds Agreement on DACA, Border Security," NBC News online, last modified January 11, 2018.
60. Josh Dawsey, Robert Costa, and Ashley Parker, "Inside the Tense, Profane White House Meeting on Immigration," *Washington Post*, January 15, 2018.
61. Ed O'Keefe, "A New Bipartisan Immigration Plan Surfaces in the Senate—and Trump Labels It a 'Total Waste of Time,'" *Washington Post*, February 5, 2018.
62. Donald J. Trump (@RealDonaldTrump), "Any deal on DACA that does not include STRONG border security and the desperately needed WALL is a total waste of time. March 5th is rapidly approaching and the Dems seem not to care about DACA. Make a deal!," Twitter, February 5, 2018, 9:36 A.M.
63. Linda Qiu, "Fact-Checking Trump's Rally in Missouri," *New York Times*, November 1, 2018.
64. Dara Lind and P. R. Lockhart, "The Diversity Visa Donald Trump Hates, Explained," Vox, last modified December 15, 2017.
65. Benjy Sarlin, "Which Side Is Donald Trump on in the Fight over Legal Immigration?," NBC News online, last modified February 7, 2017.
66. Sahil Kapur, "Trump Calls for More Legal Immigration After Pushing to Cut It," Bloomberg, last modified February 6, 2019.
67. Steve Benen, "After Demanding a Deal, Trump Rejects Another Immigration Compromise," MSNBC online, last modified February 15, 2018.
68. Donald Trump, interview by Jeanine Pirro, *Justice with Judge Jeanine*, Fox News, February 24, 2018.
69. Alvin Chang, Dara Lind, and Dylan Scott, "Every Senate Immigration Proposal on the Table, in One Simple Chart," Vox, last modified February 14, 2018.
70. James Hohmann, "The Daily 202: Trump's DACA Tweetstorm Speaks Volumes About His Presidency," *Washington Post*, April 2, 2018.
71. Pete Williams, "In Blow to Trump, Supreme Court Won't Hear Appeal of DACA Ruling," NBC News online, last modified February 26, 2018.

72. White House, "Remarks by President Trump at a California Sanctuary State Roundtable," news release, May 16, 2108.
73. Tanvi Misra, "Trump's Family Separation Policy Amplified Children's Trauma," *Roll Call*, last modified September 5, 2019.
74. Gabe Gutierrez and Adiel Kaplan, "Border Patrol Moves 100 Children to 'Appalling' Texas Facility," NBC News online, last modified June 25, 2019.
75. Tom K. Wong, "Do Family Separation and Detention Deter Immigration?," Center for American Progress online, last modified July 24, 2018.
76. Catherine Rampell, "Republicans' Inhumanity at the Border Reveals Their Grand Scam," *Washington Post*, May 28, 2018.
77. Adam Serwer, "The Cruelty Is the Point," *The Atlantic*, last modified October 3, 2018.
78. Julia Ainsley, "Top U.S. Officials Say They Got No Warning of Family Separation Policy," NBC News online, last modified July 31, 2018.
79. Ricardo Alonso-Zaldivar, "GAO: Agencies Blindsided by Trump Admin's Family Separation Policy," Talking Points Memo, last modified October 24, 2018.
80. Julia Ainsley, "Trump Admin Lost Track of Parents of 38 Young Migrant Children," NBC News online, last modified July 6, 2018.
81. Jeremy Stahl, "The Trump Administration Was Warned Separation Would Be Horrific for Children, Did It Anyway," Slate, last modified July 31, 2018.
82. William Cummings, "Trump Defends Conditions for Detained Migrant Kids, Blames Obama for Family Separations; Fact Checkers Call Foul," *USA Today*, June 23, 2018.
83. Adam Edelman, "Sessions Cites Bible in Defense of Breaking Up Families, Blames Migrant Parents," NBC News online, last modified, June 14, 2018.
84. Peter Baker, "Leading Republicans Join Democrats in Pushing Trump to Halt Family Separations," *New York Times*, June 17, 2018.
85. John Bacon, "Amid Outrage, Homeland Security Chief Kirstjen Nielsen 'Will Not Apologize' for Separating Families," *USA Today*, June 18, 2018.
86. J. M. Rieger, "The Trump Administration Changed Its Story on Family Separation No Fewer Than 14 Times Before Ending the Policy," *Washington Post*, June 20, 2018.
87. Jonathan Swan and Mike Allen, "Pressure Grows on Trump to Change Border Policy for Kids," Axios, last modified June 18, 2018.

88. White House, "Remarks by President Trump at a Meeting with the National Space Council and Signing of Space Policy Directive-3," news release, June 18, 2018.
89. Thomas Kaplan and Sheryl Gay Stolberg, "House Immigration Bill, Pitched as Compromise, Tilts to a Harder Line," *New York Times*, June 14, 2018.
90. David Nakamura, Nick Miroff, and Josh Dawsey, "Trump Signs Order Ending His Policy of Separating Families at the Border, but Reprieve May Be Temporary," *Washington Post*, June 20, 2018.
91. Maggie Haberman (@maggieNYT), "People in White House sound as confused as everyone else about what happens after the president's EO, which McGahn was against him signing. There's inter-agency disagreement about what it means. Trump now essentially tunes his chief of staff out, Trump driving the train.," Twitter, June 21, 2018, 10:14 P.M.
92. "Trump's Herky-Jerky Immigration Moves Sow Confusion," Matthew Nussbaum, Eliana Johnson and Nancy Cook, Politico, last modified June 21, 2018.
93. Nakamura, Miroff, and Dawsey, "Trump Signs Order."
94. Matthew Yglesias, "The Border Crisis Is a Reminder That Trump Has No Idea What He's Doing," Vox, last modified June 21, 2018.
95. Niels Lesniewski, "Wilbur Ross Doesn't Understand Why Furloughed Federal Workers Need Food Banks," *Roll Call*, last modified January 24, 2019.
96. White House, "Remarks by President Trump on the National Security and Humanitarian Crisis on our Southern Border," news release, February 15, 2019.
97. Bryan Pietsch, "Pentagon Pulls Funds for Military Schools, Daycare to Pay for Trump's Border Wall," Reuters, last modified September 4, 2019.
98. Andrew Exum (@exumam), "I once repurposed some excess funds we had earmarked for Ebola-related contingencies to buy blankets and cooking fuel for refugees fleeing the Islamic State and had to get four separate congressional committees to sign off on it before I could spend a dollar.," Twitter, September 4, 2019, 5:39 A.M.
99. Brian Bennett, "'My Whole Life Is a Bet.' Inside President Trump's Gamble on an Untested Re-Election Strategy," *Time*, last modified June 20, 2019.
100. Donald J. Trump (@RealDonaldTrump), "....through their country and our Southern Border. Mexico has for many years made a fortune off of the U.S., far greater than Border Costs. If Mexico doesn't immediately stop

ALL illegal immigration coming into the United States throug our Southern Border, I will be CLOSING.....," Twitter, March 29, 2019, 11:37 A.M.

101. Hallie Jackson, Jacob Soboroff, Geoff Bennett, and Alex Johnson, "Retaliating Against Democrats, Trump Says He's Considering Sending Migrants to Sanctuary Cities," NBC News online, last modified April 11, 2019.
102. Rachael Bade and Nick Miroff, "White House Proposed Releasing Immigrant Detainees in Sanctuary Cities, Targeting Political Foes," *Washington Post*, April 11, 2019.
103. Donald J. Trump (@RealDonaldTrump), "Due to the fact that Democrats are unwilling to change our very dangerous immigration laws, we are indeed, as reported, giving strong considerations to placing Illegal Immigrants in Sanctuary Cities only....," Twitter, April 12, 2019, 12:38 P.M.
104. Maria Sacchetti, "Trump's Plan to Send Migrant Detainees to Sanctuary Cities Draws Concerns About Cost, Legality," *Washington Post*, April 12, 2019.
105. Carla Herreria, "Trump Declares 'Absolute Right' to Send Undocumented Immigrants to Sanctuary Cities," HuffPost, last modified April 13, 2019.
106. Max Boot, "Guess What? Trump Has No Clue How to Stop Undocumented Immigrants," *Washington Post*, April 15, 2019.
107. Rebecca Falconer, "Trump Tells Wisconsin Rally: We're Sending Migrants to Sanctuary Cities," Axios, last modified April 28, 2019.
108. Zolan Kanno-Youngs, Maggie Haberman, Michael D. Shear, and Eric Schmitt, "Kirstjen Nielsen Resigns as Trump's Homeland Security Secretary," *New York Times*, April 7, 2019.
109. Jake Tapper, "Trump Pushed to Close El Paso Border, Told Admin Officials to Resume Family Separations and Agents Not to Admit Migrants," CNN online, last modified April 9, 2019.
110. Jake Tapper, "Trump pushed to close El Paso border, told admin officials to resume family separations and agents not to admit migrants," CNN, April 9, 2019.
111. Jake Tapper, "Trump Told CBP Head He'd Pardon Him If He Were Sent to Jail for Violating Immigration Law," CNN online, last modified April 13, 2019.
112. Eileen Sullivan and Michael D. Shear, "Trump Sees an Obstacle to Getting His Way on Immigration: His Own Officials," *New York Times*, April 14, 2019.
113. Nick Miroff and Josh Dawsey, "'Take the Land': President Trump Wants a

Border Wall. He Wants It Black. And He Wants It by Election Day," *Washington Post*, August 27, 2019.

CHAPTER 7: "We Stand by the Numbers": The Federal Budget

1. Peter R. Orszag, *The Bush Tax Cut Is Now About the Same Size as the Reagan Tax Cuts* (Washington, DC: Center on Budget and Policy Priorities, April 19, 2001).
2. Lori Montgomery, "Among GOP, Anti-Tax Orthodoxy Runs Deep," *Washington Post*, June 5, 2011.
3. Lori Montgomery, "Tax Pledge Is a Target as Deficits, Debt Grow," *Washington Post*, August 29, 2009.
4. Matt Corley, "Hatch Admits Hypocrisy: 'A Lot of Things Weren't Paid For' When Republicans Ran Congress During Bush Years," ThinkProgress, last modified January 22, 2010.
5. Victoria McGrane and Mike Allen, "GOP Gloves off for Budget Brawl," Politico, last modified March 26, 2009.
6. Elana Schor, "House GOP Unveils Its Budget Shiny Packet of Goals," Talking Points Memo, last modified March 26, 2009.
7. David Goodman and Brian Knowlton, "G.O.P. Senators Say Some Big Banks Can Be Allowed to Fail," *New York Times*, March 8, 2009.
8. Barbara Barrett, "'Beaver' Earmark in Budget Draws Attention," McClatchy, last modified March 6, 2009.
9. Barrett, "'Beaver' Earmark in Budget."
10. Mike Allen, "Why Washington Is Broken?," Politico, last modified January 26, 2010.
11. Ali Frick, "Sanford: It would Be 'Fiscal Child Abuse' to Accept Millions of Stimulus Dollars for Education," ThinkProgress, last modified March 31, 2009.
12. Mark Landler and Carl Hulse, "Still 'Far Apart' on Debt, 2 Sides Will Seek Broader Cuts," *New York Times*, July 7, 2011.
13. Ezra Klein, "Fool Me with a Budget Commission Once, Shame on You. Fool Me Eight Times . . . ," *Washington Post*, October 4, 2013.
14. Andrew Kaczynski, "Paul Ryan's Ayn Rand Moment," BuzzFeed, last modified September 21, 2012.
15. Robert Greenstein and Richard Kogan, *The Ryan-Sununu Social Security Plan: "Solving" the Long-Term Social Security Shortfall by Raiding the Rest of the Budget* (Washington, DC: Center on Budget and Policy Priorities, April 26, 2005).

16. Peter H. Wehner, "Memo on Social Security," *Wall Street Journal*, January 5, 2005.
17. Jonathan Chait, "The Legendary Paul Ryan," *New York*, last modified April 27, 2012.
18. Paul N. Van de Water, *The Ryan Budget's Radical Priorities* (Washington, DC: Center on Budget and Policy Priorities, July 7, 2010).
19. Paul Krugman, "Flim Flammed," *New York Times*, April 29, 2011.
20. Jonathan Cohn, "Would GOP Budget Actually Reduce the Deficit? (Updated)," *New Republic*, last modified April 29, 2011.
21. Paul Krugman, "The Flimflam Man," *New York Times*, August 5, 2010.
22. Bruce Bartlett, "The Balanced Budget Amendment Delusion," *New York Times*, November 15, 2011.
23. Ezra Klein, "What We Have Here Is a Failure to Communicate," *Washington Post*, March 1, 2013.
24. Jonathan Strong, "Boehner Slaps NRCC Chairman's Wrist in Chained CPI Spat," *Roll Call*, last modified April 11, 2013.
25. Rebecca Shabad, "GOP Estate Tax Repeal Would Add $269B to Deficits, CBO Says," *The Hill*, April 8, 2015.
26. *Budgetary and Economic Effects of Repealing the Affordable Care Act* (Washington, DC: Congressional Budget Office, June 2015).
27. "Now That It's Budget Time, Republicans Are Pretending to Care About the Deficit Again," Danny Vinik, *New Republic*, last modified January 29, 2015.
28. Bob Woodward and Robert Costa, "Transcript: Donald Trump Interview with Bob Woodward and Robert Costa," *Washington Post*, April 2, 2016.
29. Kelsey Snell and David Weigel, "Conservatives Ready to Support $1 Trillion Hole in the Budget," *Washington Post*, January 5, 2017.
30. White House, "Press Briefing on the FY2018 Budget," news release, May 23, 2017.
31. Nick Timiraos, "Trump Team's Growth Forecasts Far Rosier Than Those of CBO, Private Economists," *Wall Street Journal*, February 17, 2017.
32. Michael Grunwald, "Trump's Budget Scam," Politico, last modified May 23, 2017.
33. Lawrence H. Summers, "Larry Summers: Trump's Budget Is Simply Ludicrous," *Washington Post*, May 23, 2017.
34. Matthew Yglesias, "The Dumb Accounting Error At the Heart of Trump's Budget," Vox, last modified May 23, 2017.

35. Kate Kelly, Rachel Abrams, and Alan Rappeport, "Trump Is Said to Abandon Contentious Border Tax on Imports," *New York Times*, April 25, 2017.
36. "Gregg: 'This Country Will Go Bankrupt,'" CNN online, last modified March 22, 2009.
37. Judd Gregg, "Judd Gregg: Rowing into the Debt Storm," *The Hill*, February 1, 2016.
38. Max Abelson, "Trump's Tax Cuts Delight Wall Street as Debt Worries Fade," *Bloomberg Businessweek*, August 24, 2017.
39. Randall Lane, "Inside Trump's Head: An Exclusive Interview with the President, and the Single Theory That Explains Everything," *Forbes*, last modified October 10, 2017.
40. Philip Bump, "Trump Blames the National Debt on Foreign Aid as He Pushes a Tax Plan That Would Raise the Deficit," *Washington Post*, October 10, 2017.
41. Paul Krugman (@paulkrugman), "No idea in economics has been as thoroughly tested—and as completely rejected—as the notion that tax cuts pay for themselves. The reason it has been tested so much is that Republicans keep insisting that it's true, and base policy on the claim 2/," Twitter, June 12, 2019, 8:09 A.M.
42. Jim Tankersley, "It's Official: The Trump Tax Cuts Didn't Pay for Themselves in Year One," *New York Times*, January 11, 2019.
43. Dylan Matthews and Alex Ward, "4 Winners and 2 Losers from the 2019 State of the Union," Vox, last modified February 5, 2019.
44. Michael Grunwald, "Mick the Knife," Politico, September/October 2017.
45. David M. Herszenhorn, "G.O.P. Bloc Presses Leaders to Slash Even More," *New York Times*, January 20, 2011.
46. "CNBC Transcript: CNBC's Kelly Evans Speaks with National Economic Council Director Larry Kudlow at CNBC's Capital Exchange Event Today," CNBC online, last modified July 9, 2019.
47. Luis Sanchez, "Kudlow Confronted over Attacks on Obama Deficit After Rejecting CBO's Projections on Trump Budget," *The Hill*, April 9, 2018.

CHAPTER 8: **Life and Death in the Culture Wars:** Gun Control, Civil Rights, Reproductive Rights

1. Editorial Board, "It's Time to Repeal the Dickey Amendment," *Connecticut Law Tribune*, last modified September 24, 2019.
2. James Densley and Jillian Peterson, "Opinion: We Analyzed 53 Years of

Mass Shooting Data. Attacks Aren't Just Increasing, They're Getting Deadlier," *Los Angeles Times*, September 1, 2019.

3. Veronica Stracqualursi, "Congress Agrees to Millions in Gun Violence Research for the First Time in Decades," CNN online, last modified December 17, 2019.
4. Jillian Rayfield, "Mitch McConnell Email: They're Coming for Your Guns," Salon, last modified January 22, 2013.
5. Scott Clement, "90 Percent of Americans Want Expanded Background Checks on Guns. Why Isn't This a Political Slam Dunk?," *Washington Post*, April 3, 2013.
6. Greg Sargent, "ACLU: Toomey-Manchin Bill Would Make National Gun Registry Less Likely," *Washington Post*, April 15, 2013.
7. Amanda Terkel, "Pat Toomey: Background Checks Died Because GOP Didn't Want to Help Obama," HuffPost, last modified May 1, 2013.
8. "Transcript: Read the Full Text of the Second Republican Debate," *Time*, last modified September 18, 2015.
9. Jack Holmes, "Marco Rubio Wants to Know Why Nobody's Talking About 'Bomb Control,'" *Esquire*, last modified December 4, 2015.
10. Hunter, "In Wake of Mass Shooting, Marco Rubio Wonders Why We're Not Talking About 'Bomb Control' Instead," Daily Kos, last modified December 4, 2015.
11. Mark Hensch, "Trump: Obama Mulling How to 'Take Your Guns Away,'" *The Hill*, October 20, 2015.
12. Chuck Grassley interview, Capitol Hill corridor, MSNBC, February 15, 2018.
13. Associated Press, "Congress Blocks Rule Barring Mentally Impaired from Guns," last modified February 15, 2017.
14. Tara Golshan, "Republican Senator Blames the Culture of "Sanctuary Cities" for Mass Shootings," Vox, last modified October 3, 2017.
15. CAP Action (@CAPAction), "Senator John Thune says that Americans have to protect themselves from #gunviolence, adding 'get small.,'" Twitter, October 4, 2017, 9:15 A.M.
16. Ali Vitali, "Trump Says Texas Church Shooting 'Isn't a Guns Situation,' Blames Mental Health," NBC News online, last modified November 5, 2017.
17. White House, "Remarks by President Trump on the Mass Shootings in Texas and Ohio," news release, August 5, 2019.

18. "Republican Senator Says Guns Don't Kill People, Video Games Do," YouTube video, 0:22, uploaded January 30, 2013 by dkostv, https://www.youtube.com/watch?v=JuiDBr0WnZU.
19. Paul McLeod, "Trump Says Video Games Cause Violence, but Research Shows They Actually Do the Opposite," BuzzFeed, last modified May 3, 2018.
20. Phil Helsel, "Trump Blasts FBI over Parkland Shooting, Says 'Too Much Time' Spent on Russia Probe," NBC News online, last modified February 18, 2018.
21. Jonathan Swan, Margaret Talev, "Scoop: Trump Suggested Nuking Hurricanes To Stop Them From Hitting U.S.," Axios, August 25, 2019
22. Donald J. Trump (@RealDonaldTrump), "....immediately fire back if a savage sicko came to a school with bad intentions. Highly trained teachers would also serve as a deterrent to the cowards that do this. Far more assets at much less cost than guards. A 'gun free' school is a magnet for bad people. ATTACKS WOULD END!," Twitter, February 22, 2018, 7:40 A.M.
23. Philip Bump, "The Economics of Arming America's Schools," *Washington Post*, February 22, 2018.
24. Rachel Wolfe, "Trump: 10 to 20% of Teachers Are 'Very Gun-Adept.' Reality: Not Even Close," Vox, last modified February 25, 2018.
25. Lauren Fox (@FoxReports), "GOP Senate aide tells me after Trump meeting: 'It feels very much like nobody briefed the President before this meeting, which ended up being a grab-bag of Democrat priorities, instead of an effort to bring people together on proposals that actually have momentum.,'" Twitter, February 28, 2018, 4:17 P.M.
26. Haley Byrd, "Republicans Gobsmacked by Trump's Gun Control Comments," *Washington Examiner*, last modified February 28, 2018.
27. Ashley Parker, Josh Dawsey, and Ed O'Keefe, "'Negotiating with Jell-O': How Trump's Shifting Positions Fueled the Rush to a Shutdown," *Washington Post*, January 20, 2018.
28. Aaron Blake, "Sarah Huckabee Sanders Clarifies: Trump Said Lots of Stuff This Week He May Not Mean," *Washington Post*, March 2, 2018.
29. Daniel Politi, "Trump Tells NRA: Gun Rights Will 'Never Ever Be Under Siege as Long as I Am Your President,'" Slate, last modified May 5, 2018.
30. Donald J. Trump (@RealDonaldTrump), "We cannot let those killed in El Paso, Texas, and Dayton, Ohio, die in vain. Likewise for those so seriously wounded. We can never forget them, and those many who came before

them. Republicans and Democrats must come together and get strong background checks, perhaps marrying....," Twitter, August 5, 2019, 6:54 A.M.

31. Elaina Plott, "Trump's Phone Calls with Wayne LaPierre Reveal NRA's Influence," *The Atlantic*, last modified August 20, 2019.
32. Maggie Haberman, Annie Karni, and Danny Hakim, "N.R.A. Gets Results on Gun Laws in One Phone Call with Trump," *New York Times*, August 20, 2019.
33. Josh Dawsey and David Nakamura, "Trump Again Appears to Back Away from Gun Background Checks," *Washington Post*, August 20, 2019.
34. John McCain, interview by Ana Maria Cox, Air America Radio, June 11, 2009.
35. Thom Shanker and Patrick Healy, "A New Push to Roll Back 'Don't Ask, Don't Tell'," *New York Times*, November 30, 2007.
36. Roll Call Vote, 111th Congress, 2nd Session, U.S. Senate, December 18, 2010.
37. Maya Kosoff, "Twitter Explodes over Trump's Sudden Ban on Transgender Troops," *Vanity Fair*, last modified July 26, 2017.
38. Associated Press online, "Pentagon Not Aware of Trump Ban on Transgender People Serving in Military," last modified July 26, 2017.
39. Aaron Blake, "Trump's Own Defense Department Directly Contradicts His Claim About Transgender Troops," *Washington Post*, June 6, 2019.
40. "Rick Perry's Guide to Abstinence," Cindy Casares, *Texas Observer*, August 24, 2011.
41. Elspeth Reeve, "Kyl's 'Not Intended to Be a Factual Statement' Also Not a Statement," *The Atlantic*, last modified April 22, 2011.
42. Zachary Roth, "Congressman Uses Misleading Graph to Smear Planned Parenthood," MSNBC online, last modified September 29, 2015.
43. Sam Favate, "Pennsylvania Postpones Debate on Abortion Ultrasound Bill," *Wall Street Journal*, March 13, 2012.
44. Deborah Yetter, "Kentucky Law Requiring Ultrasounds Before Abortions Will Stand, Federal Appeals Court Says," *Louisville (K.Y.) Courier Journal*, June 28, 2019.
45. Luke Johnson, "Scott Walker Says He'll Sign Mandatory Ultrasound Bill," HuffPost, last modified June 12, 2013.
46. Burgess Everett, "Reid hits McConnell on gender equity," Politico, July 15, 2014.

47. Laura Bassett, "GOP Offers Birth Control Bill That Literally Does Nothing," HuffPost, last modified July 15, 2014.
48. "Republicans Aim to Flip *Roe v. Wade* with New Alabama Abortion Law," *The Rachel Maddow Show*, MSNBC, May 15, 2019.
49. Christina Cauterucci, "Ignorance Is Blessed," Slate, last modified May 15, 2019.
50. Emily Shugerman, "Legislator Pushing Abortion Ban in Alabama Says He's Not 'Smart Enough to Be Pregnant,'" Daily Beast, last modified May 14, 2019.
51. Dana Milbank, "Conservatives' Junk Science Is Having Real Consequences," *Washington Post*, May 17, 2019.
52. Gabby Orr, "Trump's Silence on Alabama Abortion Bill Is Golden for Activists," Politico, last modified May 17, 2019.
53. Walter Smith-Randolph, "Ohio Bill Would Prevent Insurance for Paying for Abortions, Limit Birth Control Coverage," WKRC online, last modified May 8, 2019.

CHAPTER 9: "Governing by Near-Death Experience": Government Shutdowns and Debt-Ceiling Crises

1. Lisa Rein and Reuben Fischer-Baum, "Hundreds of Thousands of Federal Employees Are Working Without Pay," *Washington Post*, January 23, 2019.
2. Dylan Matthews, "Here Is Every Previous Government Shutdown, Why They Happened and How They Ended," *Washington Post*, September 25, 2013.
3. Karen Tumulty, "Shutdown Crisis Shows Washington Breakdown," *Washington Post*, September 28, 2013.
4. Carl Hulse and David M. Herszenhorn, "E.P.A. and Public Broadcasting Are on House Republicans' List for Deep Cuts," *New York Times*, February 11, 2011.
5. Paul Kane and Rosalind S. Helderman, "Senate Agrees to Deal That Would Avert Government Shutdown," *Washington Post*, September 26, 2011.
6. Erik Wasson, "Obama Warns House GOP He Won't Sign Spending Bills That Break Debt Deal," *The Hill*, April 18, 2012.
7. Sabrina Siddiqui, "Richard Burr: Mike Lee Government Shutdown Threat 'Dumbest Idea I've Ever Heard Of,'" HuffPost, last modified July 25, 2013.
8. Sahil Kapur, "Senior GOPer: Try to Ditch Obamacare? Dream On, Guys," Talking Points Memo, last modified July 25, 2013.

9. Alexander Bolton, "Cruz Mocks Senate GOP Colleagues' Commitment to Defunding ObamaCare," *The Hill*, August 10, 2013.
10. Robert Costa, "Cruz to House Conservatives: Oppose Boehner," *National Review*, last modified September 27, 2013.
11. Ada Statler, "John Boehner Talks Election, Time in Office," *Stanford University (C.A.) Daily*, last modified April 28, 2016.
12. Greg Sargent, "The Morning Plum: For GOP, a Refresher on the Meaning of the Word 'Compromise,'" *Washington Post*, September 20, 2013.
13. Abby D. Phillip, "The Shutdown's Best (or Worst) Political Stunts," ABC News online, last modified October 5, 2013.
14. Sam Stein and Jason Cherkis, "The Inside Story of Why Clint Eastwood Talked to an Empty Chair at the GOP Convention," HuffPost, last modified January 26, 2016.
15. Judd Legum (@JuddLegum), "Can I burn down your house? No Just the 2nd floor? No Garage? No Let's talk about what I can burn down. No YOU AREN'T COMPROMISING!," Twitter, October 2, 2013, 2:18 P.M.
16. David M. Drucker, "GOP Stands Firm Against Funding Bill, Will Link to Debt Ceiling Fight," *Washington Examiner*, last modified October 1, 2013.
17. Keith Wagstaff, "The Government Shutdown Isn't About ObamaCare Anymore," *The Week*, last modified October 4, 2013.
18. Manu Raju, "Some Colleagues Angry with Cruz," Politico, last modified October 2, 2013.
19. Byron York, "GOP Dilemma: House Majority Failure Means Best Hope Is Weak Senate Minority," *Washington Examiner*, last modified October 16, 2013.
20. "*This Week* Transcript: Roundtable," *This Week*, ABC News, January 2, 2011.
21. Pat Garofalo, "GOP Freshman: 'Most of Us Agreed' That Raising the Debt Ceiling 'Would Be a Betrayal,'" ThinkProgress, last modified November 22, 2010.
22. Kendra Marr, "Graham: Not Raising Debt Ceiling Is 'Very Bad,'" Politico, last modified January 2, 2011.
23. David Jackson, "Obama Aide: Refusal to Raise Debt Ceiling Would Be 'Catastrophic,'" *USA Today*, January 2, 2011.
24. Michael O'Brien, "Boehner: Failure to Raise Debt Ceiling Spells 'Financial Disaster,'" *The Hill*, January 30, 2011.

25. Peter Schroeder, "Mcconnell: Obama Must Address Debt for GOP Support on Raising Debt Limit," *The Hill*, March 11, 2011.
26. Senator John Cornyn (@JohnCornyn), "Debt ceiling vote is ultimate leverage to get fiscal reform," Twitter, March 12, 2011, 11:22 P.M.
27. Kevin Bogardus, "Business Groups Likely to Align with President in Debt-Ceiling Fight," *The Hill*, April 20, 2011.
28. Ben White, "Wall Street Warns Boehner on Debt Limit," Politico, last modified April 13, 2011.
29. "Editorial: Their Temper Tantrum," *New York Times*, June 23, 2011.
30. David Brooks, "The Mother of All No-Brainers," *New York Times*, July 4, 2011.
31. Brian Beutler, "Boehner: 'This Debt Limit Increase Is Obama's Problem,'" TalkingPoints Memo, last modified July 12, 2011.
32. Robert Greenstein, "Statement: Robert Greenstein, President, on House Speaker Boehner's New Budget Proposal," press release, Center on Budget and Policy Priorities, July 25, 2011.
33. Jennifer Steinhauer, "Debt-Ceiling Fight Raised Borrowing Costs by $1.3 Billion," *New York Times*, July 23, 2012.
34. Alex Seitz-Wald, "Mitch McConnell Vows to Hold Debt Ceiling Hostage in the Future: 'We'll Be Doing It All Over,'" ThinkProgress, last modified August 1, 2011.
35. David A. Fahrenthold, Lori Montgomery, and Paul Kane, "In Debt Deal, the Triumph of the Old Washington," *Washington Post*, August 3, 2011.
36. Mitch McConnell, "McConnell: Fiscal Cliff Deal Not Great, but It Shields Americans from Tax Hike," Yahoo! News, last modified January 3, 2013.
37. David M. Drucker, "House GOP Leaders May Delay Obamacare Fight Until Debt Ceiling Talks," *Washington Examiner*, last modified September 23, 2013.
38. Ezra Klein and Evan Soltas, "The House's Debt-Ceiling Bill Is . . . Wow," *Washington Post*, September 26, 2013.
39. "Pfeiffer: White House Is Not 'Negotiating with People with a Bomb Strapped to Their Chest,'" *The Lead with Jake Tapper*, CNN online, last modified September 26, 2013.
40. Josh Dawsey and Jake Sherman, "White House Eyes Harder Line on Shutdown Talks," Politico, last modified April 19, 2017.
41. John Harwood, "Trump's Budget Director on What's on, and off, the Table for Cuts," CNBC online, last modified, April 12, 2017.

42. David Weigel, "The GOP's Alamo," Slate, last modified October 16, 2013.
43. Leigh Ann Caldwell, "Trump Signals Willingness to Drop Border Wall Funding in Budget Standoff," NBC News online, last modified April 24, 2017.
44. Alex Shephard, "Donald Trump Is a Terrible, Terrible Negotiator," *New Republic*, last modified May 6, 2017.
45. Jamelle Bouie, "Too Late," Slate, last modified January 19, 2018.
46. David Wright, "Donald Trump: Our Country Needs a Good 'Shutdown,' Suggests Senate Rule Change," CNN online, last modified May 2, 2017.
47. James Oliphant, "Trump's Dealmaker Image Tarnished by U.S. Government Shutdown," Reuters, last modified January 20, 2018.
48. Joy Reid (@JoyAnnReid), "Ouch ... John Meacham with the hammer: 'this is what government would look like without a president.' #TrumpShutdown," Twitter, January 20, 2018, 12:09 A.M.
49. Senator Brian Schatz (@brianschatz), "I've never seen such a flawed negotiation. No one is in charge. Speaker concerned about his right flank, Senate R's waiting for POTUS, POTUS changes from moment to moment. No one is sure if they have leverage or are over a barrel. It's as bad as it looks.," Twitter, January 20, 2018, 2:26 A.M.
50. Josh Dawsey and Ashley Parker, "Trump Keeps Low Public Profile During Shutdown, but Is 'Itching' to Be Involved," *Washington Post*, January 21, 2018.
51. Kevin Liptak, "'Dealmaker in Chief' Largely Absent in Weekend Shutdown Negotiations," CNN online, last modified January 22, 2018.
52. Julie Hirschfeld Davis and Maggie Haberman, "A President Not Sure of What He Wants Complicates the Shutdown Impasse," *New York Times*, January 21, 2018.
53. Hirschfeld, Davis, and Haberman, "A President Not Sure."
54. David Jackson, "Donald Trump Accepts GOP Nomination, Says 'I Alone Can Fix' System," *USA Today*, July 21, 2016.
55. Christina Wilkie, "Trump, the Master Salesman, Is Trying to Sell America a 'Democrat Shutdown'—but He Already Owns It," CNBC online, last modified December 21, 2018.
56. Josh Dawsey and Seung Min Kim, "Trump's Barometer for Success Following Russia Investigation's End—TV Ratings," *Washington Post*, March 30, 2019.
57. Seung Min Kim, Erica Werner, and Josh Dawsey, "Top Republicans Strug-

gle to Persuade Trump Not to Shut Down the Government," *Washington Post*, December 17, 2018.

58. Julie Hirschfeld Davis and Michael Tackett, "Trump and Democrats Dig in After Talks to Reopen Government Go Nowhere," *New York Times*, January 2, 2019.
59. Nicole Lafond, "Cornyn: 'If There Is' a Plan to Eschew Shutdown, 'I'm Not Aware of It,'" Talking Points Memo, last modified December 18, 2018.
60. John T. Bennett, "Trump Tweets of Talks with Dems, Invites None to Talks," *Roll Call*, last modified December 22, 2018.
61. Kathy Ehrich Dowd, "President Trump Said Federal Workers Support the Shutdown. Not True, Say Unions Representing Hundreds of Thousands," *Time*, December 26, 2018.
62. Anna Palmer, Jake Sherman, and Daniel Lippman, "Politico Playbook: Did Trump Blow the Shutdown?," Politico, last modified December 29, 2018.
63. Matthew Yglesias, "Trump's Hannity Interview Reveals a President out of Touch with Reality," Vox, last modified January 11, 2019.
64. Rebecca Ballhaus, Alex Leary, and Michael C. Bender, "Trump Considers Declaring National Emergency to Build Border Wall," *Wall Street Journal*, January 4, 2019.
65. J. M. Rieger, "For Years, Trump Promised to Build a Wall from Concrete. Now He Says It Will Be Built from Steel," *Washington Post*, January 7, 2019.
66. Jonathan Allen, "Shutdown Showdown: Pelosi Says 'No,' Trump Says 'Bye-Bye,'" NBC News online, last modified January 9, 2019.
67. Donald J. Trump (@RealDonaldTrump), "....I do have a plan on the Shutdown. But to understand that plan you would have to understand the fact that I won the election, and I promised safety and security for the American people. Part of that promise was a Wall at the Southern Border. Elections have consequences!," Twitter January 12, 2019, 11:07 A.M.
68. Jacqueline Alemany, "Power Up: Trump and Democrats Are in Parallel Universes When It Comes to the Border," *Washington Post*, January 17, 2019.
69. Sarah Jones, "Donald Trump Has No Idea How Grocery Stores Work," *New York*, January 24, 2019.
70. Teddy Kulmala, "Trump Giving In on Wall Funding 'Probably the End of His Presidency,' S.C.'s Graham Says," *State* (Columbia, S.C.), last modified January 3, 2019.
71. Erik Wasson, "House Passes Latest Plan to Fund the Government: Shutdown Update," Bloomberg, last modified January 16, 2019.

72. Mark Landler and Simon Romero, "Trump Takes Border Wall Fight to El Paso; Beto O'Rourke Fires Back," *New York Times*, February 11, 2019.
73. Niv Elis, "Senate DHS Bill Includes $1.6 Billion for 'Fencing' on Border," *The Hill*, June 19, 2018.

CHAPTER 10: "It's Like These Guys Take Pride in Being Ignorant": The Eternal Campaign

1. Tom Allison, "Gingrich Repeatedly Mischaracterized Obama's Energy Policy," Media Matters for America, last modified August 1, 2008.
2. "Monday Morning Funnies," *Wall Street Journal*, August 4, 2008.
3. Caren Bohan, "Obama Says Republican Tire-Gauge Gag Is 'Ignorant,'" Reuters, last modified August 5, 2008.
4. Dan Balz, "The Partisan Rancor Beneath Boehner's Rhetoric," *Washington Post*, July 4, 2010.
5. Alex Seitz-Wald, "Peter King: Republicans Shouldn't 'Lay Out a Complete Agenda,' Because It Might Become 'a Campaign Issue,'" ThinkProgress, last modified July 16, 2010.
6. Jordan Fabian, "Bennett Says GOP Short on Ideas," *The Hill*, June 30, 2010.
7. Brian Beutler, "Ron Johnson Freezes When Asked for His Plan to Help Middle Class," Talking Points Memo, last modified October 20, 2010.
8. Brian Beutler, "Ron Johnson on Homeless Vets: This Election Isn't About Details," Talking Points Memo, last modified October 21, 2010.
9. "Postcards from the Pledge," *The Daily Show with Jon Stewart*, Comedy Central, September 23, 2010.
10. Dan Farber, "Boehner: Americans Don't Need to Talk about Solutions Now," CBS News online, last modified September 27, 2010.
11. Scott Keyes, "GOP House Candidate Mike Kelly: 'There's Stuff to Be Cut. What Is It? I Can't Tell You,'" Talking Points Memo, last modified October 23, 2010.
12. Ed Rogers, "Two Contests for the GOP Nomination," *Washington Post*, October 24, 2011.
13. Glenn Kessler, "Euthanasia in the Netherlands: Rick Santorum's Bogus Statistics," *Washington Post*, February 22, 2012.
14. Sarah Huisenga, "Cain Says He's Ready for Questions About 'Ubeki-beki-beki-beki-stan-stan,'" CBS News online, last modified October 10, 2011.
15. Michael Memoli, "Herman Cain: 'We Need a Leader, Not a Reader,'" *Los Angeles Times*, November 18, 2011.

16. Peter Carril, "Herman Cain's Three-Page Rule," MSNBC online, last modified June 8, 2011.
17. "Stupid Answers to Stupid Questions," *American Prospect*, last modified October 3, 2011.
18. Steve Benen, "Chronicling Mitt's Mendacity, Vol. XLI," MSNBC online, last modified November 2, 2012.
19. Cadie Thompson, "Obama's 'Horses and Bayonets' Goes Viral," CNBC online, last modified November 8, 2012.
20. Katrina Trinko "Romney: New Tax Plan Can't Be Scored," *National Review*, March 7, 2012.
21. Ezra Klein, "Mitt Romney Doesn't Have a Tax Plan," *Washington Post*, August 25, 2011.
22. Jonathan Martin and Alexander Burns, "Mitt Romney's No-Policy Problem," Politico, last modified June 24, 2012.
23. Scot Lehigh, "Where Are Mitt Romney's Details?," *Boston Globe*, June 27, 2012.
24. "Senate Candidate Scott Brown: 'Do I Have the Best Credentials? Probably Not. Cause, You Know, Whatever,'" *The Week*, last modified March 24, 2014.
25. "Scott Brown on Energy: I'm Not Gonna Talk About Whether We're Going to Do Something in the Future," YouTube video, 0:48, uploaded October 21, 2014 by Jamie Henderson.
26. Phil Gingrey, "Gingrey Demands CDC Analyze Migrant Disease Threat," press release, July 8, 2014.
27. Ahiza Garcia, "GOP Rep.: Central American Immigrants May Be Carrying Ebola," Talking Points Memo, last modified August 4, 2014.
28. Eric Hananoki, "Conservatives Find a Way to Attack Obama for Fighting Ebola," Media Matters for America, last modified September 17, 2014.
29. Sally Kohn, "Ebola Scare-Mongerer Rand Paul Wants You to Think You're Going to Die," Daily Beast, last modified October 12, 2014.
30. Ashley Killough and Justin Peligri, "Rand Paul: Ebola Is "Not Like AIDS,'" CNN online, October 17, 2014.
31. Rosie Gray, "Rand Paul Says Decision on Presidential Run Will Wait Until the Spring," BuzzFeed, last modified October 23, 2014.
32. Andrew Kaczynski, "GOP Senator: ISIS Using Ebola Is a 'Real and Present Danger,'" BuzzFeed, last modified October 16, 2014.
33. Kaczynski, "GOP Senator: ISIS Using Ebola."

34. Andrew Kaczynski, "GOP Congressman: Hamas Could Infect Themselves with Ebola and Come to America," BuzzFeed, last modified October 17, 2014.
35. Evan McMorris-Santoro, "Scott Brown: 'We Would Not Be Worrying About Ebola Right Now' If Romney Won," BuzzFeed, last modified October 17, 2014.
36. Andrew Kaczynski, "Peter King Slams Doctors on Ebola, Suggests Ebola Went Airborne," BuzzFeed, last modified October 23, 2014.
37. David Weigel, "How a Reporter Got Joni Ernst to Wonder Whether the President Cares About Ebola," Bloomberg, last modified November 3, 2014.
38. Donald J. Trump (@RealDonaldTrump), "Ebola has been confirmed in N.Y.C., with officials frantically trying to find all of the people and things he had contact with. Obama's fault," Twitter, October 23, 2014, 10:24 P.M.
39. @RealDonaldTrump, October 23, 2014, 7:38 P.M.
40. Jennifer Haberkorn, "Rep. Broun: Send Money, Stop Ebola," Politico, last modified October 17, 2014.
41. Sahil Kapur, "Marco Rubio Announces Bill to Impose Travel Ban over Ebola," Talking Points Memo, last modified October 20, 2014.
42. White House Office of the Press Secretary, "Remarks by the President [Barack Obama] on American Health Care Workers Fighting Ebola," news release, October 29, 2014.
43. Kasie Hunt, "Rick Perry: Presidency Is 'Not an IQ Test,'" MSNBC online, last modified December 11, 2014.
44. David Weigel and Michael C. Bender "Jeb Bush, Confronted by DREAMer, Compares Obama Orders to Decrees of 'Latin American Dictator,'" Bloomberg News, last modified March 7, 2015.
45. Jenna Johnson, "For Carly Fiorina, Just 'Press One' to Solve the Nation's Problems," *Washington Post*, August 20, 2015.
46. Christopher Massie, "Huckabee: Dred Scott Decision 'Remains to This Day the Law of the Land,'" BuzzFeed, last modified September 10, 2015.
47. Benjy Sarlin, "2016 GOPers Woo Conservatives Ahead of South Carolina Primary," MSNBC online, last modified May 9, 2015.
48. Charles P. Pierce, "Marco Rubio in the Wilderness of Rakes, Cont'd: A Doctrine in the House," *Esquire*, last modified May 15, 2015.
49. "Walker Urges People to Visit Website for Policy Details, Which Aren't There," Talking Points Memo, last modified August 10, 2015.

50. Darren Samuelsohn, "What Do They Stand For? Good Luck Finding Out," Politico, last modified August 5, 2015.
51. Kendall Breitman, "Donald Trump Gets Personal in Attacks on Marco Rubio," Bloomberg, last modified November 2, 2015.
52. Nolan D. Mccaskil, "Trump: I'd Pick Justices Who Would Look at Clinton's Email Scandal," Politico, last modified March 30, 2016.
53. Darren Samuelsohn and Ben White, "Trump's Empty Administration," Politico, last modified May 9, 2016.
54. Zeke J. Miller, "Mismatch 2016: Donald Trump and Hillary Clinton Have Two Very Different Ideas of What It Will Take to Win the White House," *Time* online, last modified June 7, 2016.
55. Marc Fisher, "Donald Trump Doesn't Read Much. Being President Probably Wouldn't Change That," *Washington Post*, July 17, 2016.
56. Fisher, "Donald Trump Doesn't Read Much."
57. Michael Grunwald, "Donald Trump's One Unbreakable Policy: Skip the Details," Politico, last modified July 17, 2016.
58. Ethan Siegel, "Newt Gingrich Exemplifies Just How Unscientific America Is," *Forbes*, last modified August 5, 2016.
59. Siegel, "Newt Gingrich Exemplifies."
60. Josh Rogin, "Inside the Collapse of Trump's D.C. Policy Shop," *Washington Post*, September 8, 2016.
61. Andrew Kaczynski and Nathaniel Meyersohn, "Trump National Policy Adviser: Voters Don't Care About Policy Specifics," BuzzFeed, last modified September 28, 2016.
62. Brian Stelter (@brianstelter), "Here's the stat I just shared on TV: Trump's site has 9,000 words of policy proposals. Clinton's site: 112,735 words http://bigstory.ap.org/urn:publicid:ap.org:428681f59b0c40e9af1f973dd30bc1ca," Twitter, September 4, 2016, 11:20 A.M.
63. Chris Hayes (@chrislhayes), "But ultimately a Trump Presidency is a complete and total black box. No one, probably not even Trump knows what the hell it looks like. 8/8," Twitter, September 6, 2016, 9:33 A.M.
64. Stephen Ohlemacher and Marcy Gordon, "Senate Passes GOP Tax Bill, Setting Stage for Final House Vote on Wednesday," Associated Press, last modified December 20, 2017.
65. Sahil Kapur and Joshua Green, "Internal GOP Poll: 'We've Lost the Messaging Battle' on Tax Cuts," Bloomberg, last modified September 20, 2018.
66. Arthur Delaney and Matt Fuller, "Republicans Line Up Tax Cut Messaging

Bill While Abandoning Tax Cut Message," HuffPost, last modified September 12, 2018.

67. Ramesh Ponnuru, "What Republicans Aren't Telling Us in the Midterms," Bloomberg, last modified October 19, 2018.
68. Mike DeBonis, "Pelosi Is the Star of GOP Attack Ads, Worrying Democrats Upbeat About Midterms," *Washington Post*, August 9, 2018.
69. Deirdre Shesgreen and Eliza Collins, "Exclusive: Immigration Dominates GOP Candidates' TV Ads in House Contests Across the Country," *USA Today*, May 29, 2018.
70. White House, "Remarks by President Trump Before Marine One Departure," news release, October 22, 2018.
71. White House, "Remarks by President Trump and Amir Sheikh Sabah Al-Ahmed Al-Jaber Al-Sabah of the State of Kuwait Before Bilateral Meeting," news release, September 5, 2018.
72. Julia Ainsley and Daniella Silva, "Five Myths About the Honduran Caravan Debunked," NBC News online, last modified October 22, 2018.
73. Will Sommer, Lachlan Markay, Asawin Suebsaeng, and Sam Stein, "Trump's Own Team Knows His Caravan Claims Are Bullshit," Daily Beast, last modified October 23, 2018.
74. Donald J. Trump (@RealDonaldTrump), "Republicans are doing so well in early voting, and at the polls, and now this 'Bomb' stuff happens and the momentum greatly slows - news not talking politics. Very unfortunate, what is going on. Republicans, go out and vote!," Twitter, October 26, 2018, 10:19 A.M.
75. Alan Rappeport and Jim Tankersley, "Another Tax Cut? Trump and Republicans Offer a Midterm Pitch, if Not a Plan," *New York Times*, October 21, 2018.
76. Nancy Cook and Ben White, "Trump's Mystery Tax Cut Puzzles Washington," Politico, last modified October 22, 2018.
77. Alex Leary and Richard Rubin, "Follow the Bouncing Ball for Elusive Details on Trump Tax Plan," *Wall Street Journal*, October 24, 2018.
78. Laura Davison, "Mnuchin Backs Off Trump's Promise of 10% Middle-Class Tax Cut," Bloomberg, last modified December 18, 2018.
79. Peter Nicholas, "Donald Trump's Never-Ending Campaign Keeps Getting Angrier," *The Atlantic* online, last modified April 3, 2019.
80. Jason Campbell (@JasonSCampbell), "Dan Bongino praises Trump's speech:

'He absolutely blistered Hillary Clinton' (It's June 18, 2019. The 2016 presidential election was 952 days ago)," Twitter, June 18, 2019, 9:58 P.M.

81. John T. Bennett, "President Trump Can't Stop Slamming His Reelection Campaign Team," *Roll Call*, last modified July 18, 2019.
82. White House, "Memorandum of Telephone Conversation with President Zelenskyy of Ukraine," declassified by order of the president, September 24, 2019.
83. Lili Loofbourow, "The Impeachment Hearings Have Shown the Policy Costs of Trump's Narcissism," Slate, last modified November 21, 2019.
84. Michael Macagnone and Patrick Kelley, "Most Republicans on Impeachment Committees Aren't Showing Up, Transcripts Reveal," *Roll Call* online, last modified November 5, 2019.

CHAPTER 11: **Bridging the "Wonk Gap":** The Road Ahead

1. Tanya Somanader, "'We Were Strangers Once, Too': The President [Barack Obama] Announces New Steps on Immigration," White House, news release, November 20, 2014.
2. Julie Hirschfeld Davis, "Trump Backs Air Traffic Control Privatization," *New York Times*, June 5, 2017.
3. Andy Pasztor, "Senate Commerce Committee Punts on Call to Restructure Air-Traffic-Control System," *Wall Street Journal*, June 29, 2017.
4. Dan Merica, "Trump Turns to Once-Mocked Executive Orders to Tout Wins," CNN online, last modified April 27, 2017.
5. "Executive Orders," American Presidency Project, University of California Santa Barbara, last modified January 20, 2020.
6. Noah Bierman, "What's Behind All Those Executive Orders Trump Loves to Sign? Not Much," *Los Angeles Times*, March 27, 2019.
7. Bierman, "What's Behind All Those Executive Orders."
8. Annie Karni and Eliana Johnson, "'It Looks All-American': Trump Wants the Whole Package in Supreme Court Nominee," Politico, last modified July 4, 2018.
9. Jim Acosta, Juana Summers, and Eli Watkins, "WH Official: Jackson Vouching for Trump's Health on TV Helped Him Get VA Job," CNN online, last modified March 28, 2018.
10. John Wagner, "Trump Acknowledges Ronny Jackson Might Not Have Been Qualified to Lead VA," *Washington Post*, October 19, 2018.

11. White House, "Remarks by President Trump in Meeting with the National Governors Association," news release, February 27, 2017.
12. Ben Zimmer, "'Central Casting': Hollywood Lingo for Characters That Look the Part," *Wall Street Journal*, last modified March 1, 2019.
13. Ashley Parker and Maggie Haberman, "High in Tower, Trump Reads, Tweets, and Plans," *New York Times*, November 19, 2016.
14. Burgess Everett and Marianne Levine, "Republicans Whistle Past the 'Legislative Graveyard,'" Politico, last modified June 10, 2019.
15. Emily Cochrane and Catie Edmondson, "Tariff Threats Aside, the Senate Is Where Action Goes to Die," *New York Times*, June 6, 2019.
16. Camila Domonoske, "Supreme Court Declines Republican Bid to Revive North Carolina Voter ID Law," National Public Radio, May 15, 2017.
17. Paul Glastris and Haley Sweetland Edwards, "The Big Lobotomy," *Washington Monthly*, June–August 2014.
18. Paul Singer and Jarrad Saffren, "House Press Offices Expand as Other Staffs Shrink," *USA Today*, July 1, 2014.
19. Jake Sherman, "Ryan Hires 8 Communications Staffers," Politico, last modified November 2, 2015.
20. "I Was Too Hard on Mike Pence, and I'm Sorry," Matthew Yglesias, Vox, last modified July 15, 2016.
21. Molly Ball, "The Fall of the Heritage Foundation and the Death of Republican Ideas," *The Atlantic*, last modified September 25, 2013.
22. Jessica Taylor, "GOP Senator: Heritage in Danger of Not Amounting to 'Anything Anymore,'" MSNBC online, last modified October 17, 2013.
23. Paul Krugman, "The Wonk Gap," *New York Times*, September 8, 2013.
24. Karen Tumulty and Juliet Eilperin, "Trump Pressured Park Service to Find Proof for His Claims About Inauguration Crowd," *Washington Post*, January 26, 2017.
25. Joe Concha, "Trump to Jim Acosta: 'Ask the Angel Moms' If There's a Border Crisis," Politico, last modified February 15, 2019.
26. Maggie Haberman, "A President Who Believes He Is Entitled to His Own Facts," *New York Times*, October 18, 2018.
27. Duncan J. Watts and David M. Rothschild, "Don't Blame the Election on Fake News. Blame It on the Media," *Columbia Journalism Review* online, last modified December 5, 2017.